'I am energized by the ways in which this collection contributes to emerging conversations specifically in the Blue Humanities and also in the Environmental Humanities more generally. The multi-disciplinary approach taken by this collection opens the Blue Humanities to more dynamic approaches, which is a necessary task as the field continues to expand in scope. This book works to bring diverse methodological and disciplinary thinking to bear on how we understand the role of "blue" within an Australian context, but its insights extend globally and across all Blue Humanities.'

Sid Dobrin, *Professor and Chair in the Department of English, University of Florida, USA*

'From the cold Southern Ocean to tropical reefs, precarious inland waterways and the blue within art, education and digital spaces, this book dives deep into the cultural analysis of Australian waters with impressive literary elegance and analytic scope, finding the blue woven through the most pressing concerns of our times.'

Kate Judith, *Senior Lecturer at the University of Southern Queensland, Australia*

AF759807

CRITICAL APPROACHES TO THE AUSTRALIAN BLUE HUMANITIES

This interdisciplinary edited collection explores and analyses the field of the blue humanities through an Australian lens. The blue humanities is a way of understanding humanity's relationship with water and manifestations of what is referred to as the 'blue' – reefs, oceans, rivers, creeks, basins, and inland bodies of water.

In its scope, this collection emphasises both the importance of the local and the interconnectedness of Australia with global environmental concerns. It considers how we conceptualise watery spaces and shades of blue in a country where water is often marked by its absence, its ephemerality, its politicisation, and its dangers. Contributors from environmental history, environmental social science, political science, literary studies, creative arts, Indigenous Knowledge, education, and anthropology tackle various entanglements between the human, the more-than-human, and watery Australian spaces in modern culture. It is the first volume to offer a specific, dedicated focus on the intersections between Australian space and the blue humanities, and it offers a pathway for those wishing to explore, critique, and advance ideas around the blue humanities in both research and teaching.

Directly contributing to a growing interdisciplinary field, this is the first book to comprehensively examine the blue in Australia, appealing to scholars, educators, and students working across the humanities and social sciences with an interest in the environmental humanities, ecopolitics, ecocriticism, the blue humanities, cultural geography, environmental history, and the role of place.

Maxine Newlands, (PhD), is Director of the Blue Humanities Lab in Australia, and holds two adjunct research fellowships with the University of Queensland and the Cairns Institute at James Cook University. Maxine's research specialises in the advancement of novel science, in politics, policy, and marine governance.

Claire Hansen is a senior lecturer in English at the Australian National University (ANU). She is co-chair of the Blue Humanities Lab, the Heart of the Matter project, and the ANU Health Humanities Network. She is an award-winning educator, a researcher on the Shakespeare Reloaded project, co-editor of *Reimagining Shakespeare Education* (2023), and author of *Shakespeare and Place-Based Learning* (2023).

Routledge Environmental Humanities

Series editors: **Scott Slovic** (University of Idaho, USA), **Joni Adamson** (Arizona State University, USA) and **Yuki Masami** (Aoyama Gakuin University, Japan)

Editorial Board

Christina Alt, St Andrews University, UK
Alison Bashford, University of New South Wales, Australia
Peter Coates, University of Bristol, UK
Thom van Dooren, University of Sydney, Australia
Georgina Endfield, University of Liverpool, UK
Jodi Frawley, University of Western Australia, Australia
Andrea Gaynor, The University of Western Australia, Australia
Christina Gerhardt, University of Hawai'i at Mānoa, USA
Tom Lynch, University of Nebraska, Lincoln, USA
Iain McCalman, Australian Catholic University, Australia
Jennifer Newell, Australian Museum, Sydney, Australia
Simon Pooley, Imperial College London, UK
Sandra Swart, Stellenbosch University, South Africa
Ann Waltner, University of Minnesota, USA
Jessica Weir, University of Western Sydney, Australia

International Advisory Board

William Beinart, University of Oxford, UK
Jane Carruthers, University of South Africa, Pretoria, South Africa
Dipesh Chakrabarty, University of Chicago, USA
Paul Holm, Trinity College, Dublin, Republic of Ireland
Shen Hou, Renmin University of China, Beijing, China
Rob Nixon, Princeton University, Princeton NJ, USA
Pauline Phemister, Institute of Advanced Studies in the Humanities, University of Edinburgh, UK
Sverker Sorlin, KTH Environmental Humanities Laboratory, Royal Institute of Technology, Stockholm, Sweden
Helmuth Trischler, Deutsches Museum, Munich and Co-Director, Rachel Carson Centre, Ludwig-Maximilians-Universität, Germany
Mary Evelyn Tucker, Yale University, USA
Kirsten Wehner, University of London, UK

The *Routledge Environmental Humanities* series is an original and inspiring venture recognising that today's world agricultural and water crises, ocean pollution and resource depletion, global warming from greenhouse gases, urban sprawl, overpopulation, food insecurity and environmental justice are all *crises of culture*.

The reality of understanding and finding adaptive solutions to our present and future environmental challenges has shifted the epicenter of environmental studies away from an exclusively scientific and technological framework to one that depends on the human-focused disciplines and ideas of the humanities and allied social sciences.

We thus welcome book proposals from all humanities and social sciences disciplines for an inclusive and interdisciplinary series. We favour manuscripts aimed at an international readership and written in a lively and accessible style. The readership comprises scholars and students from the humanities and social sciences and thoughtful readers concerned about the human dimensions of environmental change.

For more information about this series, please visit: www.routledge.com/Routledge-Environmental-Humanities/book-series/REH

CRITICAL APPROACHES TO THE AUSTRALIAN BLUE HUMANITIES

Edited by Maxine Newlands and Claire Hansen

Routledge
Taylor & Francis Group
LONDON AND NEW YORK

earthscan
from Routledge

Designed cover image: zetter © Getty

First published 2025
by Routledge
4 Park Square, Milton Park, Abingdon, Oxon OX14 4RN

and by Routledge
605 Third Avenue, New York, NY 10158

Routledge is an imprint of the Taylor & Francis Group, an informa business

© 2025 selection and editorial matter, Maxine Newlands and Claire Hansen; individual chapters, the contributors

The right of Maxine Newlands and Claire Hansen to be identified as the authors of the editorial material, and of the authors for their individual chapters, has been asserted in accordance with sections 77 and 78 of the Copyright, Designs and Patents Act 1988.

With the exception of Chapter 2, no part of this book may be reprinted or reproduced or utilised in any form or by any electronic, mechanical, or other means, now known or hereafter invented, including photocopying and recording, or in any information storage or retrieval system, without permission in writing from the publishers.

Chapter 2 of this book is freely available as a downloadable Open Access PDF at http://www.taylorfrancis.com under a Creative Commons Attribution-Non Commercial-No Derivatives (CC-BY-NC-ND) 4.0 license.

Trademark notice: Product or corporate names may be trademarks or registered trademarks, and are used only for identification and explanation without intent to infringe.

British Library Cataloguing-in-Publication Data
A catalogue record for this book is available from the British Library

Library of Congress Cataloging-in-Publication Data
Names: Newlands, Maxine, editor. | Hansen, Claire, editor.
Title: Critical approaches to the Australian blue humanities / edited by Maxine Newlands and Claire Hansen.
Description: New York : Routledge, 2025. | Series: Routledge environmental humanities | Includes bibliographical references and index.
Identifiers: LCCN 2024006705 (print) | LCCN 2024006706 (ebook) | ISBN 9781032430492 (hbk) | ISBN 9781032430454 (pbk) | ISBN 9781003365501 (ebk)
Subjects: LCSH: Human ecology and the humanities—Australia. | Bodies of water—Australia. | Human beings—Effect of environment on—Australia.
Classification: LCC GF801 .C75 2025 (print) | LCC GF801 (ebook) | DDC 304.20994—dc23/eng/20240612
LC record available at https://lccn.loc.gov/2024006705
LC ebook record available at https://lccn.loc.gov/2024006706

ISBN: 978-1-032-43049-2 (hbk)
ISBN: 978-1-032-43045-4 (pbk)
ISBN: 978-1-003-36550-1 (ebk)

DOI: 10.4324/9781003365501

Typeset in Times New Roman
by Apex CoVantage, LLC

This work is dedicated to current and future blue humanities scholars in advancing the field; and to my family and friends, my deepest thanks for all your support in the creation of this work – Maxine

CONTENTS

Foreword by Steve Mentz *xiv*
Acknowledgements *xviii*
List of contributors *xix*

Introduction: Approaching the Australian blue humanities – from entanglements to Sea Country 1
Claire Hansen and Maxine Newlands

PART 1
Australian identities through the blue **21**

1 Blue Country: Nurturing meaningful relationships in discontinuous environments 23
Vincent Backhaus, Nalisa Neuendorf, Lokes Brooksbank, and Tahnee Innes

2 Possessing and protecting the Southern Ocean: Connection and mediation in the Antarctic work of Douglas Mawson and Alan Villiers 35
Alessandro Antonello

3 Writing the more-than-human history of northern Australia's many waters: Environmental history, the blue humanities, and the challenge of entanglement 46
Claire Brennan

4 The colour of water 58
Mia-Francesca Jones

PART 2
Sea Country, Blue Country: From the postcolonial blue to the great ocean **67**

5 Sanitary citizenship in the settler colonial city: Race, health, and hygiene in interwar urban Australia 69
Ruth A. Morgan

6 'From the viewpoint of their native element': Diving in the colonial undersea 82
Killian Quigley

7 The 'blue turn' in contemporary art: Assembling blue methods of research-creation 93
Jacqueline Chlanda, Léuli Eshrāghi, and Peta Rake

PART 3
Mediating the blue **109**

8 Ecopolitics and ecological imperialism: Activists, artisans, and the Save the Reef campaign 111
Maxine Newlands

9 Digital blues: Sense of self and the human–nature–technology connection in Australian aquatic environments 123
Melusine Martin

10 'A dancing creature of crimson and yellow': Writing the Great Barrier Reef 134
Jessica White

PART 4
Beyond the anthropocentric blue **147**

11 Moving waters, muddy edges: Ibis in Brisbane 149
Gillian Paxton

12 A whale of a journey: On the connectivity between pygmy blue whales in Indonesia, Australia, and beyond 162
Putu Liza Kusuma Mustika

13 HMS *Pandora* and the sea: Tracing eighteenth-century Polynesian artefacts and their entanglement with the Pacific Ocean 172
Jasmin I. Günther

PART 5
Imagining blue futures **185**

14 Colourblindness in/of place: Memory, colonial place, and education's ignorance of the blue 187
Bryan Smith

15 Eco-art and reeling in Anthropogenic adversity 198
Robyn Glade-Wright

16 Waves of cognition: Towards an Australian Blue Shakespeare ecosystem 211
Alys Daroy, Joshua Zeunert, and Rahul K. Gairola

Index *222*

FOREWORD

Australian blue humanities: A global view

Seen from the northern hemisphere, across oceans and the equatorial line, the island continent of Australia appears a far-off land between waters, with its borders lapped by the Pacific, Indian, and Southern Oceans. Where I am writing, on the calm waters of Long Island Sound not far from New York City, Australia's ocean cultures appear distant. The northeastern United States has neither coral reefs nor tropical rainforests, though in some ways the surf cultures of California and Florida bear family resemblances to those on the coasts of Australia. Reading across that planetary distance, the great pleasure in encountering this collection of essays has been discovering the ways in which an Australian blue humanities appears distinctly *Australian*. Its geophysical environments and ecological systems make Australia radically different from other parts of global Anglophone culture. Along with Australia's combination of connection to and distance from the global North, the nation's recent history of fossil fuel extraction, cultural responses to Indigenous activism, and water cultures of many different kinds make the coasts, reefs, and watery margins of Australia ripe for innovation in the blue humanities. We who think, teach, and write – as well as swim, sail, and surf! – in this mode have much to learn from these pages.

In this brief introduction, I will suggest how a view from the Global North and Atlantic waters might productively intermingle with these distinctive Australian voices. My opening term, borrowed from the introduction, will be 'entanglement'. Building on ecological and feminist analyses by scholars such as Stacy Alaimo and Astrida Neimanis, the concept of entanglement brings together both physical proximity and metaphorical resonance. As Neimanis influentially observes, humans are all 'bodies of water', and we share aquacorporeality with other living things, larger bodies such as rivers, lakes, and oceans, and the planetary system as a whole

(2017). The implications of Neimanis's assertion, especially when combined with concepts such as Alaimo's 'trans-corporeality' and newer insights by Melody Jue, John Durham Peters, and others who treat seawater as media, place the watery blue at the centre of human relationships with our more-than-human environment (Alaimo, 2010; Jue, 2020; Peters, 2015). To recognise our entanglement with water in its multiple forms, from the liquid in our bodies to the vapour in the air, represents an enabling step in blue humanities thinking. When we have eyes to see, blue appears everywhere – even if much of the water with which we find ourselves entangled appears less clear flowing blue than muddy brown, biotic green, gleaming white ice, or many other colours.

The authors of this book's incisive chapters begin from a basic recognition of water's infiltrating touch. For me as an American scholar, the special force of these essays surfaces the varied currents of the blue humanities and the ways that water entangles itself inside and alongside cultural objects. These essays also speak to some larger-scale conversations in global water-infused thought that I will outline briefly here.

Global connections

We know that 'oceans connect' on local and global scales, and the human history of the southern continent includes many stages of global seafaring, starting with the original settlement of Australians perhaps 65,000 or more years ago.[1] Modern ideas of a global or World Ocean have been central to recent constellations of human interconnectivity, resonantly visualised by the 'Blue Marble' photograph of the planet Earth taken by the Apollo 17 crew from space in 1972. Blue humanities scholars have troubled in meaningful ways the burst of optimism that, for some environmentalists, accompanied the Blue Marble image. Elizabeth DeLoughrey's *Allegories of the Anthropocene* and her continuing argument for a 'critical ocean studies', which she distinguishes against a broader 'blue humanities', argues for water-thinking that rejects neoliberal economic fantasies (2019, 2022).[2] Sidney Dobrin's *Blue Ecocriticism and the Oceanic Imperative* also assumes a global perspective, in particular on the 'protein economy' generated by the fishing industry (2021). The view from Australia can productively invert northern-centric visions of globalisation, just as attention to the undersea ecologies of shipwrecks and their encrustations, in recent work by Sara Rich, Natali Pearson, and Killian Quigley, can reveal new perspectives (Rich, 2021; Pearson, 2022; Quigley, 2023). These scholars and the authors of this book can help reorient understandings of oceans as globalising vectors.

Black and blue as historical matrices

The cruel history of the North Atlantic Slave Trade has massively shaped and still distorts the politics, population dynamics, and cultures of the communities that border these waters. Recent scholarship has expanded ideas about the Black Atlantic into a global vision that places the Black and Indigenous experience of the Middle Passage at the centre of cultural and environmental history. Adding an Australian perspective to this developing conversation can be illuminating. The slave trade in the North Atlantic remains distinct from the colonial settlement of Australia, though some of the same ships, humans, nations, and environmental forces were at work in both. It might be valuable to entwine some of the ideas made visible in Brad Pascoe's *Dark Emu*, among other works of Australian Critical Indigenous Studies, with the works of American Black Studies scholars such as Christina Sharpe, Dionne Brand, and Alexis Pauline Gumbs (Pascoe, 2018; Sharpe, 2016; Brand, 2001; Gumbs, 2020). The voices who are reorienting oceanic histories in Anglo-Caribbean contexts will make potent interlocutors for the Australian blue humanities.

Blue as capitalist horror

Many scholars who explore global waters and Anthropocene environments have focused on capitalism's half-millennia of exploitation as key to making sense of oceanic modernity. The collection, *Anthropocene or Capitalocene?*, edited by Jason Moore, provides a helpful commingling of key voices in this debate, including the influential ecotheorist Donna Haraway (2016). Perspectives from Australia can usefully supplement Moore's eco-globalism and other perspectives that have dominated debates about the Anthropocene and the many competing 'cenes, from Plantationocene to Capitalocene to Pyrocene, that have been proposed.[3] Historical analyses of capitalism, colonialism, and environmental history, especially those written by scholars in the global North, have a tendency to focus on Northern communities. The story of waterborne capitalism comes with a different valance from the point of view of the Southern hemisphere and Australia in particular.

Towards Anthropocene water

Juxtaposing the competing voices of North and South seems necessary to advance the blue humanities as a global current of scholarly and artistic communication. Keeping those geographic strands to a degree separate may be hard to avoid. But the larger project, ultimately, aims to bring into community all manners of local and global bodies of water. To quote Alice Te Punga Somerville's inspiring words, which also appear in this volume's introduction, 'I want to ask if Ocean Studies might be better understood if it were itself an ocean: without a singular starting point or origin; endlessly circulating' (2017). This vision of dynamic flow and non-hierarchical circulation motivates a global blue humanities that aims to bring multiple ideas, places, and forms into dialogue. As the Australian-born scholar James L. Smith observes about medieval water cultures, there is a 'hydro-social' intersection between water and human cultures that seems important to explore (2018). A broad vision of Anthropocene water in the present and hydro-social forces operating throughout human and nonhuman history can enable new ways of thinking with and alongside watery forms. That dynamic flow characterises the best possible futures for this fast-moving discourse. The essays in this volume represent an important contribution to global blue humanities. They also issue an important challenge to scholars and scholarly institutions that reside mostly in the global North. To know planetary water, we must flow with all its currents.

Steve Mentz
St John's University
New York City

Notes

1 I am grateful to my hosts on my most recent trip to Australia in 2019 for encouraging me to read *Dark Emu* (Pascoe, 2018).
2 A developing critique of the blue humanities appears in DeLoughrey (2022).
3 A short, and already out-of-date, round-up of roughly two dozen proposed 'cenes appears in Mentz (2019).

References

Alaimo, S. (2010). *Bodily Natures: Science, Environment, and the Material Self.* Bloomington: Indiana University Press.
Brand, D. (2001). *A Map to the Door of No Return: Notes on Belonging.* Toronto: Doubleday Canada.
DeLoughrey, E. (2019). *Allegories of the Anthropocene.* Durham: Duke University Press.
DeLoughrey, E. (2022). Mining the seas: Speculative fictions and futures. In I. Braverman (Ed.), *Laws of the Sea: Interdisciplinary Currents* (pp. 144–163). London: Routledge.
Dobrin, S. (2021). *Blue Ecocriticism and the Oceanic Imperative.* London: Routledge.
Gumbs, A. P. (2020). *Undrowned: Black Feminist Lessons from Marine Mammals.* Chico: AK Press.
Jue, M. (2020). *Wild Blue Media: Thinking Through Seawater.* Durham: Duke University Press.
Mentz, S. (2019). *Break Up the Anthropocene.* Minneapolis: University of Minnesota Press.
Moore, J. (Ed.). (2016). *Anthropocene or Capitalocene? Nature, History, and the Crisis of Capitalism.* Binghamton: PM Press.
Neimanis, A. (2017). *Bodies of Water: Posthuman Feminist Phenomenology.* London: Bloomsbury.
Pascoe, B. (2018). *Dark Emu: Aboriginal Australia and the Birth of Agriculture.* London: Scribe.
Pearson, P. (2022). *Belitung: The Afterlives of a Shipwreck.* Honolulu: University of Hawaii Press.
Peters, J. D. (2015). *The Marvelous Clouds: Toward a Philosophy of Elemental Media.* Chicago: University of Chicago Press.
Quigley, K. (2023). *Reading Underwater Wreckage: An Encrusting Ocean.* London: Bloomsbury.
Rich, S. (2021). *Shipwreck Hauntography: Underwater Ruins and the Uncanny.* Amsterdam: Amsterdam University Press.
Sharpe, C. (2016). *In the Wake: On Blackness and Being.* Durham: Duke University Press.
Smith, J. (2018). Medieval water energies: Philosophical, hydro-Social, and intellectual. *Open Library of the Humanities, 4*(2). doi: https://doi.org/10.16995/olh.228
Te Punga Somerville, A. (2017). Where oceans come from. *Comparative Literature, 69*(1), 25–31.

ACKNOWLEDGEMENTS

We wish to acknowledge that this book was edited on Bindal and Wulgurukaba, Gadigal, Ngunnawal, and Ngambri Country. We pay our respects to custodians and Country.

Thank you to the founding members of the Blue Humanities Lab, without whom this collection would not have been possible. We wish to thank The Cairns Institute at James Cook University and the Australian National University. Our deepest thanks to Steve Mentz for sharing his wisdom in the opening pages of this collection – we are very grateful. Thank you to the editors at Routledge, especially Rosie and Matthew, and to the series editors and reviewers of this collection for their excellent advice. Thank you to all of our contributors for sharing your blue humanities thinking in the pages that follow.

CONTRIBUTORS

Alessandro Antonello is a senior research fellow in history at Flinders University, on Kaurna Country in Adelaide. He is the author of *The Greening of Antarctica: Assembling an International Environment* (2019) and his research investigates the environmental and international histories of the Antarctic and the world's oceans in the twentieth and twenty-first centuries.

Vincent Backhaus is a research fellow in The Cairns Institute with affiliation to the Indigenous Education and Research Centre, James Cook University. Currently, he is researching co-design processes with Great Barrier Reef (GBR) Traditional Owners involved in Sea Country governance across research and management of Crown of Thorns Starfish along the GBR.

Claire Brennan is a senior lecturer in history at James Cook University, Australia. She researches the history of hunting (including crocodile hunting), and the history of northern Australia more broadly. She is currently working on the Coral Discovery project, creating a digital map of European scientific voyages entering the Pacific before 1834.

Lokes Brooksbank is a senior project manager at the Great Barrier Reef Foundation working within Traditional Owner Strategic Initiatives and communications. He recently submitted his PhD thesis which analysed governance systems in the Papua New Guinea Petroleum sector.

Jacqueline Chlanda is a researcher, writer, and arts professional based in Meanjin/Brisbane, where she is currently Education Manager at The University of Queensland (UQ) Art Museum. She has a PhD in Art History, English Literature, and Philosophy, previously worked at the National Gallery of Australia, and has taught art history at UQ and the Queensland College of Art.

Alys Daroy is Lecturer in English and Theatre at Murdoch University, Western Australia. Publications include *Shakespeare, Ecology, Adaptation with Paul Prescott* (Bloomsbury Arden Shakespeare) and various eco-Shakespeare works. Alys is Artistic Director of Australia's first professional eco-Shakespeare company, Shakespeare South, and formerly acted as Sustainability Consultant at Shakespeare's Globe.

Léuli Eshrāghi (Seumanutafa + Tautua Sāmoan, Persian, Cantonese) intervenes in display territories to prioritise global Indigenous and Asian diasporic visuality, sensual and spoken languages, and ceremonial–political practices. They are Curator of Indigenous Arts at Musée des Beaux-Arts de Montréal and Curatorial Researcher at Large at University of Queensland Art Museum. They hold a PhD in Curatorial Practice from Monash University and completed a postdoctoral fellowship at Concordia University.

Rahul K. Gairola is the Krishna Somers Senior Lecturer in English and Postcolonial Literature and Fellow of the Indo-Pacific Research Centre at Murdoch University. He has published widely, and is an area editor for *Oxford Bibliographies in Literary and Critical Theory* and a South Asian book series editor for Routledge.

Robyn Glade-Wright Robyn Glade-Wright is an arts educator and a practising contemporary environmental artist in Queensland, Australia. She has presented over fifty solo exhibitions in public and private galleries, and her large-scale environmental works of art call attention to the role humans have played in climate change, environmental pollution, and species loss.

Jasmin I. Günther finalised her PhD at James Cook University in a cotutelle arrangement with Aarhus University, Denmark. Previously, she studied Anthropology at Göttingen University in Germany and worked as an assistant curator at the Linden-Museum in Stuttgart. She has recently joined the MARKK in Hamburg for a provenance research project.

Tahnee Innes is an anthropologist and Yawuru-nation woman from Broome in Western Australia. Her research interests include Indigenous curatorship, art and ethnography, and socio-political landscapes. Currently, she works in native title with the Indigenous people of Cape York, Australia.

Mia-Francesca Jones is a creative writer and researcher living on Dja Dja Wurrung Country. Her research explores how ecological emotions like solastalgia, homesickness, and ecoanxiety manifest in personal narrative and ecoautobiography. In 2022, she was awarded the William Thomas Williams Postgraduate Scholarship for a creative scholar.

Melusine Martin aims to understand humans' relationships with nature in a digital context, focusing on the interactions between self-identity, natural environments, and digital technology. She has a PhD in Environmental Sociology from Sorbonne Université and James Cook University (Cotutelle). Her first monograph, *Reconnect with Nature* (*Reconnectez-vous à la Nature*) (2024), is published by Larousse.

Ruth A. Morgan is the Director of the Centre for Environmental History at the Australian National University. She is the award-winning author of *Running Out? Water in Western Australia* (2015, UWA Press) and co-author of *Cities in a Sunburnt Country: Water and the making of urban Australia* (2022, Cambridge University Press).

Putu Liza Kusuma Mustika (Icha) is an adjunct researcher focusing on cetacean conservation at the College of Science and Engineering at James Cook University. She leads CETASI (Cetacean Sirenian Indonesia) NGO, coordinates the online 'Whale Stranding Indonesia' database, and contributes to the IUCN Cetacean Specialist Group and International Whaling Commission.

Nalisa Neuendorf is a senior research fellow and Social Science lead at the Papua New Guinea Institute of Medical Research. As an applied social anthropologist, she has many years of research experience in the areas of culture and society, gender, and health, utilising a range of qualitative research methodologies.

Gillian Paxton is an environmental anthropologist at The Cairns Institute at James Cook University. She has a special interest in multispecies ethnography and her current work focuses on the more-than-human politics of technologically assisted ecological adaptation in the Great Barrier Reef.

Killian Quigley is a research fellow at the Australian Catholic University's Institute for Humanities and Social Sciences and Honorary Postdoctoral Fellow at the Sydney Environment Institute, University of Sydney. He is the author of *Reading Underwater Wreckage: An Encrusting Ocean* (2023, Bloomsbury) and co-editor, with Margaret Cohen, of *The Aesthetics of the Undersea* (2019, Routledge).

Peta Rake is Director/Senior Curator at University of Queensland Art Museum. She has held roles at CCA (San Francisco), Institute of Modern Art (Brisbane), Banff Centre for Arts and Creativity, Walter Phillips Gallery, and Banff International Curatorial Institute (Banff, Canada).

Bryan Smith looks at anti-racist and decolonising readings of humanities and social sciences education. His current line of work looks at the making of settler place and how everyday features of the urban landscape write settler possession into the material and symbolic spaces of communities.

Jessica White is the author of the novels *A Curious Intimacy* (2008, Penguin Random House Australia) and *Entitlement* (2012, Penguin eBooks), and a hybrid memoir about deafness, *Hearing Maud* (2019, UWA Publishing). She is co-editor of *Science Write Now*, a journal of creative writing inspired by science, and she has won awards, funding, and residencies. Jessica is currently Senior Lecturer in Creative Writing and Literature at the University of South Australia.

Joshua Zeunert is a Scientia Associate Professor at UNSW, Sydney. He specialises in sustainability, landscape architecture, and urban design. His recent DECRA research investigates large-scale Australian agricultural systems. Two of Zeunert's books are *Landscape Architecture and Environmental Sustainability* (2017, Bloomsbury) and the *Routledge Handbook of Landscape and Food* (2018, Routledge).

INTRODUCTION

Approaching the Australian blue humanities – from entanglements to Sea Country

Claire Hansen and Maxine Newlands

Australia is defined by water. From the Traditional Knowledges of Aboriginal and Torres Strait Islander peoples to the reminder in the Australian National Anthem that 'our home is girt by sea',[1] the blue shapes us. Water makes us as we make, remake, and unmake it (Neimanis, 2017, pp. 21–22). Water, in the cosmologies of Indigenous Australian peoples, is 'a sacred and elemental source and symbol of life' (Langton, 2006). Many Indigenous Australian cultures 'have a water being in their ongoing creation or dreaming stories' (Giblett, 2018, p. 148).[2]

This sacred water being from Indigenous dreaming stories is glimpsed in the opening pages of Waanyi author Alexis Wright's *Carpentaria*, proclaimed as a – or even *the* – 'great Australian novel' (Shek-Noble, 2021, p. 196).

> Picture the creative serpent, scoring deep into – scouring down through – the slippery underground of the mudflats, leaving in its wake the thunder of tunnels collapsing to form deep sunken valleys. The sea water following in the serpent's wake, swarming in a frenzy of tidal waves, soon changed colour from ocean blue to the yellow of mud. The water filled the swirling tracks to form the mighty bending rivers spread across the vast plains of the Gulf country. The serpent travelled over the marine plains, over the salt flats, through the salt dunes, past the mangrove forests and crawled inland. Then it went back to the sea. And it came out at another spot along the coastline and crawled inland and back again. When it finished creating the many rivers in its wake, it created one last river, no larger or smaller than the others, a river which offers no apologies for its discontent with people who do not know it. This is where the giant serpent continues to live deep down under the ground in a vast network of limestone aquifers. They say its being is porous;

DOI: 10.4324/9781003365501-1

> it permeates everything. It is all around in the atmosphere and is attached to the lives of the river people like skin.
>
> *(Wright, 2006, pp. 1–2)*

In its movement across the Gulf of Carpentaria (off the northern coast of Australia), the 'wet body' of this ancient serpent elegantly interweaves the elements and reinforces the connectedness of freshwater with salt. The serpent 'permeates everything' and is connected so deeply to the river people that it is 'like skin'; there can be no disentanglement of elements, no separation of the human and more-than-human, in this creation story. As Vincent Backhaus et al. evocatively articulate in the opening chapter of this collection,

> The littoral and riparian boundaries of these channels shape relationships, and carry life, water, peoples, and things – connecting places and stories as an epistemic and ontological flow of connectivity (Langton, 2006). Water becomes a Storyline that illustrates the connections between river flows that span out across the continent and the sea, becoming a conduit for meanings to connect with deep oceanic currents. These meanings return over Country as rain to replenish river flows once again.
>
> *(p. 27)*

As the editors of this collection, we recognise our privilege as white, western scholars working in Australia and emphasise that an Australian blue humanities must begin with and learn from Aboriginal and Torres Strait Islander peoples' relationships with water – as exemplified in Wright's novel. This is especially critical given that, as Jeremy Hills et al. write of the Blue Pacific context, First Nations people's knowledge of the ocean's connections to humans has either been lost or 'continues to be largely unrecognised in ocean policy and in the protection and management of the ocean and its resources' (2022, p. 274). Such knowledge, they add, is in need of 'revival and renewal' (Hills et al., 2022, p. 274). Geographer and anthropologist Marcia Langton writes of First Nations people's water knowledge:

> Bodies of water are constituted by Aboriginal people as being more than just a physical domain. They are construed spiritually, socially and jurally, according to the same fundamental principles as affiliations to terrestrial places in the land; the dialogic relationship in Indigenous thought between the ancestral past and its effect on human existence derives from the Aboriginal understanding of the transformative powers of the spiritual beings that inhabit those places, whether a landscape, waterscape or skyscape. For Aboriginal people, the distinctions between land and water are not absolute; rather, they are constituted as a cultural space that includes both land and water bodies that may be viewed as a unitary cultural phenomenon, such as a traditional estate in a littoral riparian or marine

> area, or a water body believed to be the resting place of an ancestral being and deemed to be a sacred site.
>
> *(Langton, 2006)*

Water relations are social and ontological (Langton, 2006). Natures, agencies, and bodies, as Serenella Iovino and Serpil Oppermann tell us, are 'entangled in endless ways', their more-than-human materiality 'a constant process of shared becoming' (Iovino & Oppermann, 2014, p. 1). Humans and water are not separate entities (Wallace-Wells, 2019).

The story of the Australian blue has though, since colonisation, been marked by violence, dispossession, environmental degradation, and the increasingly catastrophic impact of a changing climate. The impact of climate change is 'accelerating', with 'abrupt change' anticipated (Turner et al., 2020, p. 7). Discourses about Australian climate change tend to localise around devastating bushfires, floods, drought, and mass bleaching of corals at the Great Barrier Reef (the Reef), a 'major site of narrativity' in blue humanities thinking (Oppermann, 2023, p. 15). It is now predominantly accepted that the greatest threat to the Reef is climate change, amongst a host of related threats including temperature change, invasive species, severe weather, and water quality degradation (Walpole & Hadwen, 2022). United Nations Educational, Scientific, and Cultural Organization (UNESCO) remains concerned over the increasing local stressors from land clearing, water quality, and run-off from agricultural and pastoral activities as well as expansions in industry and coastal communities the length of the Reef. UNESCO in 2021 repeated concerns[3] about the long-term health of the Great Barrier Reef ecosystem, and a UNESCO–International Union for Conservation of Nature (IUCN) joint mission in 2022 investigated threats from local stressors, like pollution, coastal development, ocean temperature rises, and over-fishing.[4] There are also stark links between climate change impacts and exporting of fossil fuels in Australia. Coral reef experts Terry Hughes et al. have pointed out that while the significance of climate change as a threat to the Reef is recognised by the Australian Government, Australia 'is the world's largest exporter of seaborne fossil fuels, and also has the world's highest per capita emissions of greenhouse gases' (Hughes, Day, & Brodie, 2015, p. 508). The Australian Commonwealth and Queensland Governments address UNESCO's concerns through The Reef 2050 Long-Term Sustainability Plan (The Plan) as a joint strategy to ensure the long-term health of the Great Barrier Reef. Key measures to protect the Reef include the creation of no-fishing zones across one third of the World Heritage site, achievement of water quality targets by 2025, and more ambitious CO_2 emissions reduction targets.[5] These targets are in line with the national medium-term emissions reduction target for 2035 and the Queensland Government's goals for a net zero emissions target by 2030.

While the Reef is Australia's most visible symbol of the blue and of climate change impacts, there are many other water systems that are threatened by anthropogenic climate change, including Ningaloo Reef in Western Australia, the Great

Artesian Basin – which lies beneath much of the eastern half of the continent – and the Murray–Darling Basin river system, which comprises 23 rivers in the south east of Australia. As Jason Alexandra writes, 'the Anthropocene's rivers are not playing by the past's rules' (Alexandra, 2021, p. 16). Also vulnerable to climate change are groundwater recharge (McCallum et al., 2010) and trees, which in northern Australia's old growth forests are dying at a vastly accelerated rate, likely due to 'atmospheric water stress' driven by global warming (Bauman et al., 2022). Flood events are also increasingly recognised as natural events with 'human factors that have caused the hazard', and yet, as Margaret Cook reflects, 'many people still believe that Mother Nature is to blame or that floods can be prevented' (2023, p. 221). This illuminates that water is a 'climate parable: an abundant resource made scarce through government neglect and indifference, bad infrastructure and contamination, careless urbanization and development' (Wallace-Wells, 2019, p. 87).

Søren Frank reminds us that the 'Blue Humanities is born out of the climate crisis' (Frank, 2023, p. 16), and this collection identifies ways for people to better understand climate change impacts through creativity, art, history, storytelling, and disciplines from the social sciences and humanities. This book thus falls into a broader suite of environmental humanities (Cohen & Foote, 2021) scholarship that seeks to not only examine, through a historicist lens, engagements with specific aspects of our environments, but also looks ahead to the way our engagements are shaping our presents and futures, and how the humanities, arts, and (social) sciences can contribute to a better understanding of human relationships with the blue. We emphasise the critical focus of the Australian blue humanities on climate change because, in the words of Steve Mentz: 'Global climate change defines the central challenge of the current generation, and scholarship about water in all its forms will be necessary to make sense of our disrupted ecosystems' (Mentz, 2023, p. 7).

The following essays examine how the blue humanities is developing in relation to Australian waters and how it can be used to illuminate our entangled relationships with water. As Mentz writes: 'The burden of the blue humanities follows water's sinuous trails wherever they flow and traces their dampness in our environments and memories' (Mentz, 2023, p. xiv). In *Critical Approaches to the Australian Blue Humanities*, we follow water's sinuous trails across the Australian continent and beyond: from the tropical waters of Far North Queensland to the sanitation of interwar Perth, to the exploration of the Southern Ocean and the impact of salty seas on shipwreck artefacts, to the waters in which pygmy blue whales travel from Indonesia to the Antarctic and beyond.

While this is the first collection to focus exclusively and explicitly on Australian blue humanities, it is far from the first work to engage with the blue in Australian history, politics, and culture. We acknowledge that there is significant work already done – under other names and fields, from environmental humanities to cultural studies – that could be identified as Australian blue humanities scholarship. Much of the work in the pages to come recognises pre-existing 'blue' work across disciplines. The blue humanities has many related neologisms: the oceanic turn or turns (Anderson et al., 2023, pp. 3–4), hydro-criticism and ocean history (Alaimo, 2019, p. 1; Mentz, 2023, p. 1), 'critical ocean studies' (DeLoughrey, 2019, p. 29) as well as blue ecocriticism (Dobrin, 2021), 'terraqueous ecocriticism' or 'aquatic environmentalism' (Oppermann, 2019, p. 445), 'blue cultural studies' (Mentz, 2023, p. 6), 'ocean humanities' or 'thalassography' (Jue, 2020, p. 16), 'humanist oceanic studies' and 'Blue Anatolian Humanism' (Oppermann, 2023, p. 2).[6] This book therefore seeks to identify common threads across approaches, disciplines, and fields by weaving together theory, lived experience, and case studies. In so doing, we gesture to a way forwards that seeks to follow the 'swirling tracks' across Australia's mudflats and salt plains, mangrove forests and bush, rivers and droughts, floods and coral reefs, in order to identify and model critical approaches to the Australian blue humanities.

Towards entanglement: Defining the blue humanities

Like its subject matter, the blue humanities is capacious and always in flow. In her incisive criticism of the language we give to the oceans and to ocean studies, Indigenous Studies and literary studies scholar Alice Te Punga Somerville writes:

> I want to ask whether Ocean Studies might be better understood as if it were itself an ocean: without a singular starting point or origin; endlessly circulating. Not beyond genealogy, because nothing is, but possessed of a genealogy that is impossibly and beautifully wide. I attempt to sketch the ocean—and Ocean Studies—from the perspective of those who have not needed a "turn to the sea" because we were already there.
>
> *(Te Punga Somerville, 2017, p. 28)*

Te Punga Somerville astutely critiques the 'oceanic turn' as a predominantly western perspective. As Jon Anderson et al. observe:

> the ocean, it seems, has shifted from a marginal to a central concern. In some areas, though – and it is important to note – the ocean has always been central to thinking and being (Hau'ofa, 1995, 2008). It is western scholarship itself which can be accused of emptying the ocean for its own filling.
>
> *(Anderson et al., 2023, p. 4)*

For many – often marginalised – voices and perspectives, the ocean was always already central. Te Punga Somerville's call to decentre the critical gaze from certain types of ocean (a northern lens, a centring of the Atlantic, or specific understandings of the Pacific) has been taken up by blue humanities scholars James L. Smith and Mentz, who write that 'we must constantly forget how to navigate, because the tools of navigation were born in an age of exploitation' (Smith & Mentz, 2020). We agree that 'learning to inhabit pluralistic and inclusive blue humanities' (Mentz, 2023, p. 5; Smith & Mentz, 2020) is imperative, and through the blue humanities seek to comprehend the complex and intricate relationships between humans, more-than-humans and water, its cultural, historical, and emotional connections, and its role in shaping human culture, society, and individual experiences. The blue humanities illuminates the impact of human actions on marine ecosystems, the threats posed by climate change, pollution, and unsustainable practices, and aims to recognise multiple perspectives on human–water relations.

As a 'watery criticism', the blue humanities both 'describes the complex works of water' and imagines 'ways to change our relationship to it' (Mentz, 2023, p. 1). Importantly, as Oppermann articulates, the blue humanities entertains 'the propositions that the main problems and insecurities in aquatic ecosystems are ineluctably social and cultural', and thus 'the possibilities for any hopeful change are socially and culturally situated' (Oppermann, 2023, p. 1). The blue humanities thus stakes a claim for the vital role of humanities, arts, social sciences, and critical thinking in understanding and shaping our relationships with the more-than-human. As Elizabeth DeLoughrey writes of the Anthropocene, 'the conversation has been dominated by the geophysical and social sciences, which tend to privilege positivist methods and have little to say about the vitality of the arts and humanities' (DeLoughrey, 2019, p. 2). The blue humanities stresses this vitality when it comes to understanding the blue. In an Australian environmental context, such 'cross-fertilisation' is essential (Neumeier, 2020, p. 1).

The first step of the blue humanities is to defamiliarise the terrestrial in favour of 'hydro-ecological frameworks' (Oppermann, 2019, pp. 444–445). For Mentz, 'the blue humanities name an ocean-infused way to reframe our shared cultural history' (Mentz, 2020, p. xviii) and encourages 'oceanic' rather than 'terrestrial' global thinking (Mentz, 2015, p. xxi). While the blue humanities is 'dazzled by oceanic vastness' (Mentz, 2023, p. 1), its interest has expanded to encompass all forms of water (Oppermann, 2023, pp. 9–10). Mentz seeks 'a blue humanities that responds to water's material complexity and imaginative polyphony'; the field thus encompasses water 'in all its phases' – vapour, ice, fresh, and salt water (Mentz, 2021, p. 193). As Mentz – the founder of the blue humanities – indicates, the term takes its name 'from the nineteenth-century notion of "blue water" – i.e., the ocean out of sight of land' (Mentz, 2021, p. 191). It belies its own name in its expanding focus on all shades of water, because the waters of our world are not always – or perhaps not even often – blue (it is also far from the only element in our natural world which boasts this hue). As Melody Jue reminds us, 'the ocean has not always been blue', but perhaps 'blue is just the beginning' (Jue, 2020, pp. x, xii).

The blue humanities is thus multi-hued.[7] These colours of the natural world are one way to communicate more-than-human watery stories; coral reefs, writes Oppermann, 'project their stories through their dwindling colours as meaningful signs to convey their dying cries. Colour is their primary tool of expression, used with mastery to stay alive and for conveying their sad stories' (Oppermann, 2023,

p. 15). Many of the chapters in this collection engage with diverse colours and manifestations of water: Alessandro Antonello writes that the Southern Sea is not 'only blue and white' but comprises green and brown water, ice, and, for whalers in the late nineteenth century, the red of whale blood; Mia-Francesca Jones illuminates the complexity of the 'colour of water' through her examination of the movement from fresh and brackish to salt water; Claire Brennan reminds us that the Australian 'blue' is not always blue; and Bryan Smith articulates the need for a 'cyan' education. The blue humanities thus embraces many hues and moves beyond salt or fresh water (Oppermann, 2023, p. 3) to envelop what Mentz calls the 'many varieties of planetary waters' (Mentz, 2023, p. xviii).[8]

With rapid growth in the early twenty-first century, the blue humanities is defined by its transdisciplinarity: it has burst forth across disciplines including literary studies, environmental history, art, anthropology, and more (Mentz, 2023, p. 1), and includes sociocultural, literary, historical, aesthetic and ethical perspectives (Oppermann, 2023, p. 1). The methods of the blue humanities are also 'multimodal', exceeding academic writing to include 'artistic work' and 'activism' (Mentz, 2023, pp. 1, 2). Typically, scholars 'combine data provided by marine sciences with cultural and literary theories to modify our anthropocentrically oriented conceptions and discourses of planetary oceans' (Oppermann, 2023, p. 3). These methods are inherently multidisciplinary: 'Writing, thinking, and creative work that flies the flag of the blue humanities responds to water's intimate paradoxes in many modes, from poetry to literary criticism to history' (Mentz, 2023, p. xiv). While Mentz is trained in literary studies, he is careful to reaffirm the inherent transdisciplinarity of the blue humanities, and the usefulness of literature to that end. He seeks an inclusive approach that does not decentre the work of historians, ecologists, and marine biologists (Mentz, 2023, p. xv). The transdisciplinarity of the field enables the blue humanities to promote 'hybrid forms and strategies' to best confront the complex watery realities that we face (Oppermann, 2023, p. 8).

The blue humanities attends not only to representations of water, to multiple forms and scales of water, and to our relationships with water, but also to watery ways of thinking. Cecilia Chen et al. observe that 'we already think with water, both physiologically and in this use of watery language and metaphor' – water 'permeates the way we think' (2013, p. xxiii). Mentz identifies 'three fundamental features of blue humanities thinking': physical contact with water, disorientation, and the emergence of new collectivities, which operate across multiple scales, from the community and nation to posthuman networks (Mentz, 2023, p. xvi).

At its core, the blue humanities is interested in human connections with the more-than-human blue: with all forms of material blue and with how our relationships to it enmesh us in specific ways of being, knowing, and connecting. It is fundamentally driven by a view of ourselves and our world as integrally and always entangled. This introduction seeks to extend 'entanglement' as a mode of blue humanities thinking and as a framework for understanding our relationships with water, especially in the context of climate change impacts.

The concept of entanglement is based on the work of Iovino and Oppermann (Iovino & Oppermann, 2014; Oppermann, 2019); it is a concept that recurs in blue humanities scholarship.[9] In their material ecocritical approach, entanglements describe the interaction of 'material and discursive agencies at work in the world's becoming' (Iovino & Oppermann, 2014, p. 10) and the 'terraqueous entanglements of human and marine life' (Oppermann, 2019, p. 445). Such entanglements are 'at once intimate and uncanny, distant and proximal, and above all unpredictable' (Oppermann, 2019, p. 461). This concept counters the idea of nature as 'a reified thing in the distance' separate from humanity (Morton, 2010, p. 4) and reinforces the deconstruction of the nature/culture binary. As DeLoughrey notes, 'In keeping with work in feminist, postcolonial, and Indigenous studies that does not bifurcate nature and culture, the concept of ecology as it is used here always includes the human' (DeLoughrey, 2019, p. 7). Such a perspective aligns with field-defining work by thinkers including Epeli Hau'ofa. His image of Oceania is 'vast' and 'expanding', comprising humans and more-than-humans: 'Oceania is us. We are the sea, we are the ocean' (Hau'ofa, 2008, p. 62). Hau'ofa's description of Oceania's sea of islands illuminates this complex entanglement of human and ocean relationships.

By focusing on blue entanglements, we can better recognise the agency of the more-than-human. This way of thinking 'denature[s] anthropocentric narrative representations' of water and emphasises 'storied seas' and the 'affective agency' of the ocean (Oppermann, 2023, p. 35). This moves beyond anthropocentric ideas of nonhuman forms as 'passive, inert, unable to convey any independent expression of meaning' (Iovino & Oppermann, 2014, p. 2). It enables us to recognise nonhuman stories or storied water, which comprises the stories of sea turtles to coral reefs (Oppermann, 2023, pp. 13–14). From this new materialist perspective, agency does not privilege the human but recognises matter as 'agentic' (Oppermann, 2023, p. 55); agency 'is a pervasive and inbuilt property of matter, as part and parcel of its generative dynamism' (Iovino & Oppermann, 2014, p. 3).

In considering entanglement as a guiding framework for the blue humanities, we understand it to occur across three interrelated sites of interaction: storytelling, human and more-than-human relationships, and between the material and metaphorical blue – water as both 'substance' and 'poetics' (Chen et al., 2013, p. 9), comprising both 'the *discursive* and the *real*' (Oppermann, 2019, p. 446).[10]

A more comprehensive overview of the blue humanities is beyond the scope of this introduction and would return to waters already navigated: for a rich overview of blue humanities thinking and scholarship, readers are advised to seek out Mentz's *Introduction to the Blue Humanities* and Oppermann's *Blue Humanities: Storied Waterscapes in the Anthropocene* (Mentz, 2023; Oppermann, 2023). A sample of other influential work on blue theory includes Chen et al.'s *Thinking with Water*, Jue's *Wild Blue Media*, and Astrida Neimanis' *Bodies of Water* (Chen et al., 2013; Jue, 2020; Neimanis, 2017). These and other examples of 'embodied water-scholarship represent a theoretical leading edge of blue humanities scholarship in the 2020s' (Mentz, 2023, pp. 10–11).

By combining insights from the natural sciences and the humanities, this collection of essays practises an inclusive blue humanities and encourages a collective contemplation of the significance of water in shaping identities, memories, and beliefs, thereby cultivating a sense of responsibility towards conserving and preserving our precious aquatic ecosystems for future generations.

From the big wet to Sea Country: An Australian blue humanities

This collection is focused on Australian encounters with the blue. It exemplifies what Frank identifies as 'increasing interest in the maritime global south' (Frank, 2023, p. 8). The book echoes the Australian continent's blue boundary of the Global Ocean, or, as acknowledged by the Australian Government, the Pacific, Southern, and Indian Oceans, and the Timor, Tasman, and Coral Seas, the Arafura Sea, and the Southern Ocean.[11] While acknowledging this kind of anthropocentric line drawing and legal recognition of boundaries, we also note the geographical complexity of what constitutes the Australian blue. For example, the Tasman Sea separates the mainland of Australia from its southern island, Tasmania or lutruwita, and Australia's 'External Territories' include islands in the Pacific, Indian, and Southern Oceans, the Coral and Timor Seas, and several islands closer to the mainland which span over 1,000 square kilometres.[12] Geoscience Australia numbers over 8,000 islands that fall within Australian territory. The boundaries of what constitutes an Australian blue are especially complex, as Country was never ceded ('Aboriginal sovereignty', 1988).[13] As Putu Liza Kusuma Mustika explores in her chapter in this collection, national boundaries are further problematised when considering the movement of more-than-human creatures like the pygmy blue whale. Thus, while recognising that this book focuses on 'Australian' blue humanities, we also seek to illuminate the complexities of this concept. How we define the 'blue' and how we define 'Australian' blue are live questions that, we believe, should remain open, dynamic, and inclusive. This collection examines the fluidity and contingency of boundaries across its chapters.

In considering the Australian blue, we encounter extremes. Australia is famously characterised by the excesses of the wet and the dry – both colloquial Australian phrases. 'The wet' and 'the big wet' refer to the monsoonal rains common in the tropical zones, the wet season of the tropical north, or an extended period of high rainfall, while 'the big dry' refers to 'profound drought' (MacNally et al., 2014).[14] Environmental historian Rebecca Jones describes drought as 'a backdrop to life in Australia. Water – and its absence – has shaped soil, vegetation and animals, and shadows society, giving and withholding, pushing and pulling, hindering and encouraging' (2017, p. 10). Her study of long-term farm diaries illuminates human resilience and the drought as both a natural and cultural disaster (2020, p. 13).

An end to the 1870s drought is depicted as 'apocalyptic' in one diary entry, writes Jones. The entry reads: 'You could not call it rain; the water just tumbled from the sky . . . Wherever you looked great cracks had broken the ground and the rain pouring in enlarged them' (Jones, 2017, p. 28). Over a century later, in *The Dry*, a popular Australian novel and subsequent film, Jane Harper offers a vivid contemporary account of drought:

> The drought had left the flies spoiled for choice that summer. They sought out unblinking eyes and sticky wounds as the farmers of Kiewarra levelled their rifles at skinny livestock. No rain meant no feed. And no feed made for difficult decisions, as the tiny town shimmered under day after day of burning blue sky. 'It'll break,' the farmers said as the months ticked over into a second year. They repeated the words out loud to each other like a mantra, and under their breath to themselves like a prayer. But the weathermen in Melbourne disagreed. Besuited and sympathetic in air-conditioned studios, they made a passing reference most nights at six. Officially the worst conditions in a century. The weather pattern had a name, the pronunciation of which was never quite settled. El Niño.
>
> *(Harper, 2017, p. 1)*

At the opposite end of the watery spectrum, a similar extremity is seen in the exquisitely illustrated Indigenous children's book *Big Fella Rain*, which traces the interconnected, complex flow-on effects that occur when 'way up north . . . the mighty storms come, lightning flashes through liquorice clouds' (Webber & Martins, 2018). The muddy pools become streams, then creeks, the billabong comes to life, and the wildlife – from dragonflies to dingoes – is revitalised. While here, the big wet is a welcome, life-giving affirmation, the dangers of the extreme wet are also brought to life in artwork such as W. C. Piguenit's famous 'The flood in the Darling 1890' (1895) and, more recently, Blak Douglas' portrait 'Moby Dickens', which won the 2022 Archibald Prize.[15] Douglas' portrait depicts Wiradjuri artist Karla Dickens amidst the 2022 flood disaster in northern New South Wales, Australia. It references another famous 'blue' work, *Moby Dick*, and critiques the Australian Government's response to the disaster. Another Wiradjuri experience of flood is depicted in Anita Heiss' novel *Bila Yarrudhanggalangdhuray* (*River of Dreams*), which explores the 1852 Gundagai flood. While the cyclical flooding of the Murrumbidgee River is interpreted and respected by the Wiradjuri people, the recently arrived settlers struggle to survive.

> The Wiradyuri people know to move to higher ground. They have done it many times before, and they will probably do it again. Camping and having ceremony on the flood plain is their way of life, but so is understanding the ways of nature, the flow and power of the Marrambidya, and when to get out of his way.
>
> *(Heiss, 2022, p. 28)*

In documenting floods from settlement to the present in Townsville, 'Queensland's drought-prone, dry tropics' (Lloyd et al., 2020, p. 100), Rohan Lloyd et al. describe 'the contrast between drought and deluge' as a reflection of 'a 150-year struggle between the natural environment and the city in Queensland' (2020, p. 99).

This drastic movement between the extremes of 'the dry' and 'the big wet' echo across the contemporary Australian consciousness and also in blue humanities scholarship. Mentz points to the vitality of this contrast: 'The variable opposition between the wet and the dry has long been at the heart of my thinking about blue environments' (Mentz, 2023, p. xiv). He positions the wet as emphasising 'disorder, disorientation, and rupture' in juxtaposition to 'a dry countermovement that attempts to make sense and meaning out of disaster' (Mentz, 2015, p. 11). He adds:

> The contrast between wet swirl and dry structure operates a fundamental division that animates blue humanities thinking. The conceptual distinction between experience and knowledge and the felt difference between feeling and form flow from the contrast between the wet and the dry.
>
> *(Mentz, 2023, p. xiv)*

While the wet and the dry form predominant contemporary modes of thinking about Australian relationships with water (and its absence), this introduction also emphasises the primacy of Aboriginal and Torres Strait Islander peoples' perspectives which recognise that the sea, like the land, is 'alive' (Whitehouse et al., 2014, p. 56). In engaging in any kind of blue humanities scholarship, we must heed Te Punga Somerville's reminder that:

> The presence of Indigenous people across and beyond this region surely demands an Ocean Studies that imagines, seeks, and engages the intellectual work about oceans that islanders have been doing for decades—indeed, for millennia.
>
> *(Te Punga Somerville, 2017, p. 30)*

Australian 'waterscapes can be better understood through Indigenous worldviews', which see landscapes and waterscapes as 'meaningfully entangled in social relations' (Langton, 2006). This incorporates the concept of Sea or Saltwater Country, which

> is an imperfect English translation of very old, extant and now flourishing Torres Strait Islander concepts and practices concerning care for saltwater 'land' that extends beyond littoral boundaries to the continental shelf and oceanic horizons. The terms refer to coastal, island and marine environments that together constitute the traditional estates of the coastal and maritime peoples in Australia.
>
> *(Whitehouse et al., 2014, p. 58)*

Sea Country is the encompassing of land and sea scapes that are identified as Traditional lands. Across the Reef, there are 'at least 44 Traditional Owner groupings with interests in sea country' (Dale et al., 2016). In Indigenous cosmologies water and land 'are treated as inseparable, as part of one integrated "country"' (Burdon et al., 2015, p. 336). Sea Country is often spatially identified from the sources of a river out to the oceans. Sea Country comes from at least 50,000 years of cumulative knowledge about the oceans (Dale et al., 2016). This interconnectivity between land and sea is a physical and spiritual connection for First Nations peoples:

> fundamental to the way we interact with the sea is our belief that we are a part of the sea, and the sea is a part of us, always has been and always will be. This belief is maintained through our stories passed down from one generation to the next.
>
> *(Dillon, 2002, p. 13)*

Country is more than just place; it is a part of identity. But as Hilary Whitehouse et al. articulate, Sea Country was 'deeply challenged by waves of immigration and colonising legislative practices' (Whitehouse et al., 2014, p. 60). Sea Country was brought into direct conflict with the 'legal fiction' of *terra nullius*, upon which the British colonisation of Australia from 1788 was predicated (Whitehouse et al., 2014, p. 61).

> The replacement of Sea Country as lived, known, loved, understood, mapped and cared for with the notion that it was empty land, or *terra nullius*, is mind-boggling to contemporary scholastic sensibilities.
>
> *(Whitehouse et al., 2014, p. 61)*

Prior to colonisation, Saltwater Country was managed 'within strictly delineated boundaries by many different peoples from many different language groups'. But after Australian Federation in 1901, beaches were declared public spaces and Federal law excluded private property rights between high and low tide zones (Whitehouse et al., 2014, p. 60). 'While this legal manoeuvre manifested Australian ideals of freedom (the beach as open, with free passage for all), this move contradicted much older laws, lores and customs' (Whitehouse et al., 2014, p. 60).[16] Integrating the rich sources of local or Traditional Knowledge into our engagements with water is not only morally required, but it can also bring diverse benefits to research, knowledge use in decision-making and management, and conservation success. The relevance of Traditional Knowledge is particularly important to both better understand and adapt to impacts from climate change.

In collating this book, the authors are cognisant of the vexed histories that lay behind a collection that refers to an 'Australian blue'; this term should never whitewash the multi-hued, multi-perspectival complexity of the Australian continent's blue spaces, its histories, traumas, its human and more-than-human inhabitants, and the ongoing complex interactions between them.

Chapter overview

This collection is an invitation to think critically and transdisciplinarily about the ways in which the blue functions in Australian culture, history, arts, scholarship, education, and practice. This book, we hope, is a diving board into an exploration of our entanglements with the blue in Australia, and a starting point for further investigation into the Australian blue humanities.

Critical Approaches to the Australian Blue Humanities understands the blue humanities as fundamentally interested in entanglements between human and non-human entities, between the many social science and humanities disciplines, and between traditional and new forms of knowledge. For this reason, this collection interweaves multiple disciplines and diverse manifestations of the Australian blue. The book comprises five themes to focus an exploration of the Australian blue humanities: Australian identities through the blue; Sea Country and the postcolonial blue; mediating the blue; beyond the anthropocentric blue; and imagining blue futures.

While this book is focused on the Australian blue, it does so while always keeping a broader context in view. We follow DeLoughrey's intervention in these debates over the need for critical attention to the micro and macro, the local and global scales (DeLoughrey, 2019, p. 6). She proffers the island as 'an important figure for this simultaneous relationship between the part and whole, the local and global' (DeLoughrey, 2019, p. 9). For this reason, blue humanities founder Steve Mentz offers a foreword to this collection that contextualises what is in some ways a localised investigation within a global frame. Throughout, we see Australian borders as a vexed and fluid question rather than as clear, firm lines, and thus this collection troubles the distinction between localised and planetary blue.

In Part 1: 'Australian identities through the blue', the authors explore how human identity – a sense of nationhood and our individual construction of self – is shaped by the blue, and how this process also shapes blue environments. The collection begins with Vincent Backhaus and co-authors' reflection on how Blue Country is grounded in relationships between land, sea, sky, and people. Speaking as a People of Knowledge, Backhaus et al. illuminate what they call '*Blue identity*' or '*Countried identity*', which is always in a process of becoming and active in placemaking. In Chapter 2, 'Possessing and protecting the Southern Ocean', Alessandro Antonello examines the Southern Ocean as a space of Australian imagination, characterised by logics of possession and protection. He focuses on two case studies – the explorer Douglas Mawson and the whaler and writer Alan Villiers – and highlights the role of radio telegraphy in constructing the space of the Southern Ocean in the Australian imagination. Continuing with an environmental history focus, in Chapter 3 Claire Brennan writes the more-than-human history of northern Australian waters and considers the development of watery thinking in Australian environmental history. Part 1 of this collection concludes with a creative personal essay by Mia-Francesca Jones, 'The colour of water', which explores intimate familial relationships alongside the unique ecosystems of the Great Barrier Reef and Far North Queensland.[17] In this intermingling of the personal with the blue, Jones examines how individual and familial identity are constructed through engagements with blue space.

Postcolonial thinking flows throughout this collection and is visible here in the introduction, in the illuminating foreword, and in each part of the book. To gesture to its critical embeddedness in the blue humanities, this book devotes Part 2 to postcolonial scholarship while also threading it throughout the collection. This is in line with Smith and Mentz's view that:

> A key element in creating a global and multicultural blue humanities scholarship lies in listening to, and engaging with, those whose stories have been ignored. In the mostly terrestrial context of eco-studies today, one notable trend that the blue humanities would do well to follow is the rising visibility and influence of Indigenous Studies.
>
> *(Smith & Mentz, 2020)*

Part 2 of the collection is explicitly devoted to examining postcolonial analyses of blue entanglements in Australia. This works as part of what Oppermann identifies as 'resistance to the previously unquestioned ideology of western discourses by which the seas are represented in arts and literature' (Oppermann, 2023, p. 6). We begin with Ruth Morgan's historical analysis of 'hydrosocial relations' in the interwar Australian cities of Perth, Melbourne, and Darwin (p. 71). Morgan considers how freshwater was a means of governing not only space but raced, classed, and gendered bodies. Chapter 6 attends to divers and diving as 'icons of knowledge-making' in colonial waters (p. 82). Killian Quigley examines writer and political organiser Jean Devanny's sojourn to the Great Barrier Reef and queries the 'settler-colonial undersea' and 'what human presences the seascapes of naturalist and recreational diving have been more and less capable of assimilating' (p. 91). We turn to critical transcendence of colonial occupation and exploitation of blue spaces in Chapter 7 with 'The "blue turn" in contemporary art: Assembling blue methods of research-creation'. Here, Jacqueline Chlanda, Léuli Eshrāghi, and Peta Rake showcase the work of four Australian artists presented as part of *Blue Assembly*, an ongoing research and programming arc at the University of Queensland Art Museum underpinned by the concept of 'tidalectic curating'. The artistic work explored in this chapter highlights complex relationships between blue histories and colonial violence (from Blackbirding to militourism) and uses aesthetics to focus on 'relationships and responsibilities outside of empire and capital' (p. 105).

In Part 3, in a nod to Jue's focus on 'how media and mediation occur *through the ocean*' (Jue, 2020, p. 22), the authors examine diverse ways in which to 'mediate' engagements with the blue – whether that is through politics, the digital, or literature. Maxine Newlands explores the history of the 'Save the Reef' campaign in Chapter 8 as a case study of transdisciplinary activism and advocacy between the science, community, and the blue. In Chapter 9, Melusine Martin contemplates digitised blue space and proffers the concept of 'digital solastalgia', building on Glenn Albrecht's term, as a way of redefining place attachment and human relationships with blue spaces. There is a return to the Reef in Chapter 10, where Jessica White examines literary representations of the Reef in a range of genres, from the poetry of Judith Wright to Australian children's literature. Through media and politics, through digital space, and through literature, these chapters interrogate how we construct and mediate our relationships and entanglements with the blue.

To move beyond an anthropocentric gaze, Part 4 centres on the more-than-human, and in the words of Iovino and Oppermann, 'the agentic properties of material forms' (2014, p. 7). Here, the authors give voice to nonhuman perspectives in an exploration of more-than-human relationships – from watery birds to cetaceans to artefacts. In her poignant exploration of the ibis – a waterbird in contemporary urban Australia – Gillian Paxton reveals how these birds 'disrupt our delusions of separation and control' (p. 158), making visible watery flows and processes often kept hidden in our contemporary lives. In Chapter 12, Putu Liza Kusuma Mustika moves beyond the Australian continent as she takes us on a cetacean journey from the waters of Indonesia down the western coast of Australia, following the journeys of the pygmy blue whale. This reveals much about the way we define terrestrial and aquatic boundaries, while Mustika also offers a reflection on the impact of the blue on her own spiritual identity and on our limited understanding of the whales' sense of self. As Stacy Alaimo reflects, there is something of a quandary in seeking to give voice to the more-than-human in a human collection. But there is potential for 'science and humanistic (or posthumanistic) inquiries to yield something solid or lucid that could enable us to understand something of the creatures of the sea' (Alaimo, 2019, pp. 429–430). In consideration of the more-than-human, Chapter 13 (Jasmin I. Günther) focuses on eighteenth-century Polynesian artefacts recovered from the wreck of the HMS *Pandora* and traces the afterlives of these objects – especially the role of water in shaping their histories. In focusing on artefacts, shipwreck, and the entanglement of the sea with human stories and histories, Günther's work can be read through Quigley's concept of the 'encrusting ocean' and his recognition of how human and more-than-human somethings can be understood as 'growings together', encrustations that are 'reimagining marine matters' (Quigley, 2023, p. 32). Through these more-than-human stories, we recognise Oppermann's claim that solutions to the environmental crisis are to be found in 'new stories told by aquatic narrative agencies' (Oppermann, 2023, p. 57).

Part 5: 'Imagining blue futures' looks forward to imagine how we might work with and understand the blue in the future: in education, in environmental art, and in literary studies. This final section takes up Mentz's call that 'As seas rise, glaciers melt, and warmer air absorbs more moisture, the blue humanities will have much to say and do in our changing ecological conditions' (Mentz, 2023, p. xviii). The chapters of this section imagine 'possible futures for human intimacy with water' (Mentz, 2023, p. xviii). It is telling that two of these chapters advocate strongly for pedagogical transformation. In Chapter 14, Bryan Smith offers us a vision for 'cyan education' in his critique of the colourblindness of contemporary Australian education. Smith argues that we need to 'tint our critical encounters' with cyan and tease out 'the complex entanglements of the green and blue as spaces of memorial production and representation' (p. 195). Environmental artist Robyn Glade-Wright shares examples of eco-art in her chapter, exploring how we can look to the arts as a communicator of climate science. Her work exemplifies a subversive approach to anthropogenic climate change and pollution, as Glade-Wright recycles plastic and marine debris into environmental art that synthesises complex topics into emotive, tangible artefacts designed to motivate sustainable behaviour. This, like Chapter 7, demonstrates how the blue humanities brings attention 'to new art practices that highlight the fragility of oceanic habitats' (Oppermann, 2023, p. 6). 'Imagining blue futures' concludes with a focus on Shakespeare in Australia as a recognition that the 'blue humanities' was coined by Mentz, a Shakespearean literary studies scholar and historian. In the final chapter, Alys Daroy, Joshua Zeunert, and Rahul K. Gairola propose an ecosystem approach for Australian blue Shakespeare and consider postcolonial blue pedagogy in teaching Shakespeare's *The Tempest*. This chapter models interdisciplinary dialogue and points us towards ongoing learning as a key element of blue humanities thinking.

The purpose of this collection is to amplify and give space to a diverse range of voices that illuminate the dynamism and complexity of human relationships with the blue. It is guided by an 'inclusive' approach to the blue humanities (Smith & Mentz, 2020). With this in mind, this introduction returns to Wright's *Carpentaria*, which articulates diverse and complex human relationships and entanglements with the blue, incorporating the material and metaphorical blue, and storytelling that encompasses human and more-than-human perspectives. When faced with a climate catastrophe – a cyclone and flood event – one of the protagonists of *Carpentaria* survives on the sea. 'In order to know', we are told, 'he would have to become a scholar of the sea' (Wright, 2006, p. 495). To face our own uncertain futures, we too must become scholars of the sea.

Notes

1 Australian National Anthem, https://www.pmc.gov.au/honours-and-symbols/australian-national-symbols/australian-national-anthem
2 Giblett adds that the idea of the 'Rainbow Serpent' is, however, a generalisation by anthropologists. It is vital to avoid such generalisations, as Aboriginal and Torres Strait Islander peoples' cultures are 'diverse and heterogeneous' (2018, p. 148). Giblett acknowledges that he is a non-Indigenous person and so does not speak with cultural authority. Luke Taylor also identifies that the 'Rainbow Serpent' has entered 'Australian parlance' and, thanks to 'anthropological exposition', its image has enjoyed increasing appeal (1990, p. 329).

3 The Great Barrier Reef was endowed as a UNESCO world heritage site in 1981. The bestowment was given for the Outstanding Universal Value of the coral and reef system in the northern section. The 'in danger' status arose when in 2012, the then Australian Minister for the Environment, Greg Hunt, approved what would become one of the world's largest coal ports in central Queensland (Lankester et al., 2015). The Galilee Basin, when fully operational, would produce 70 million tonnes of coal, worth $1.5 billion in royalties, most of which would be shipped through the Great Barrier Reef. UNESCO raised concerns over the impacts to the Reef from increased industrialisation, port expansion, and dredging within the Great Barrier Reef Marine Park area. In a joint plan, the Australian and Queensland Governments responded with the State Party Report on the state of conservation of the Great Barrier Reef World Heritage Area in late 2013 (Tarte & Hughes, 2020). In 2015 the Reef 2050 Long-Term Sustainability Plan (Authority, 2015; 'The Reef 2050 Long-Term Sustainability Plan') was accepted by UNESCO as a programme of works, and the plan and government actions are reviewed annually. Any deviation or delay in implementing the Plan would result in an 'In Danger' listing. However, the annual bleaching of corals from climate change impacts has repeatedly raised concerns, with increasing pressure from political parties and some of the science community (Bertacchini et al., 2016; Hofman & Hughes, 2018) to list the Reef as in danger, arguing not enough is being done to address rising global temperatures.

4 'Great Barrier Reef: Australia to put in place urgent safeguarding measures requested by UNESCO', Press Release, 6 June 2023, https://www.unesco.org/en/articles/great-barrier-reef-australia-put-place-urgent-safeguarding-measures-requested-unesco#:~:text=For%20many%20years%2C%20UNESCO%20has%20sounded%20the%20alarm%20on%20the,This%20warning%20had%20global%20resonance

5 'Great Barrier Reef: Australia to put in place urgent safeguarding measures requested by UNESCO', Press Release, 6 June 2023, https://www.unesco.org/en/articles/great-barrier-reef-australia-put-place-urgent-safeguarding-measures-requested-unesco#:~:text=For%20many%20years%2C%20UNESCO%20has%20sounded%20the%20alarm%20on%20the,This%20warning%20had%20global%20resonance

6 For more on the genesis of the blue humanities, see Oppermann, 2023, p. 2.

7 For more on colour, see Cohen, 2013.

8 Oppermann notes that the blue humanities 'still prioritizes the imperiled seascapes' because of the extent of Anthropocene-related devastation affecting the World Ocean (Oppermann, 2023, p. 5).

9 Chen et al. refer to our 'material entanglements with the watery world' (2013, p. 12). Killian Quigley writes of 'the fascinating, and frequently vexing, entanglements of word and world, thought and matter, artifice and ecology' (2023, p. 8). Mentz uses the concept of entanglement to refer to types of human and more-than-human interactions (Mentz, 2023, pp. 10, 25, 32, 49, 138).

10 For critical arguments on the role of metaphor, see Oppermann, 2023, pp. 23–24, 53–54.

11 'Oceans and Seas', Geoscience Australia, 7 June 2023, https://www.ga.gov.au/scientific-topics/national-location-information/dimensions/oceans-and-seas

12 'Islands', Geoscience Australia, 7 June 2023, https://www.ga.gov.au/scientific-topics/national-location-information/landforms/islands

13 See the 2017 Uluru Statement from the Heart: https://ulurustatement.org/the-statement/view-the-statement/; https://humanrights.gov.au/our-work/aboriginal-and-torres-strait-islander-social-justice/publications/uluru-statement-heart#:~:text=The%20Uluru%20Statement%20from%20the%20Heart%20(the%20Statement)%20is%20an,Parliament%20and%20a%20Makarrata%20Commission

14 Langton observes that Aboriginal groups in the Cape York Peninsula (she includes 'the Guugu Yimithirr, the Lamalama, the Alu Aya (or Kuku Thaypan), the so-called "Koko Warra", and the people of Barrow Point, among others') recognise six seasons instead of the 'common understanding of a "wet" and a "dry" season' (2006).

15 W. C. Piguenit, 'The flood in the Darling 1890' (1895), https://www.artgallery.nsw.gov.au/collection/works/6105/#about. Blak Douglas, 'Moby Dickens' (2022), https://www.artgallery.nsw.gov.au/prizes/archibald/2022/30418/
16 See also scholarship on Aboriginal and Torres Strait Islander peoples' water rights, including Burdon et al. (2015).
17 For more examples of creative critical explorations of human relationships with water, see *Water Lore* (Roulière & Egerer, 2022).

References

'Aboriginal sovereignty – never ceded.' (1988). *Australian Historical Studies*, *23*(91), 1–2. doi: 10.1080/10314618808595786

Alaimo, S. (2019). Introduction: Science studies and the blue humanities. *Configurations*, *27*(1), 1–10.

Alexandra, J. (2021). Navigating the Anthropocene's rivers of risk – climate change and science-policy dilemmas in Australia's Murray–Darling Basin. *Climatic Change*, *165*(1–2), 1–21.

Anderson, J., Davies, A., Peters, K. A., & Steinberg, P. (2023). Introduction: Placing and situating ocean space(s). In K. A. Peters, J. Anderson, A. Davies, & P. Steinberg (Eds.), *The Routledge Handbook of Ocean Space* (pp. 3–19). Routledge.

Authority, G. B. R. M. P. (2015). *Reef 2050 Integrated Monitoring and Reporting Program*. Retrieved from https://www2.gbrmpa.gov.au/our-work/programs-and-projects/reef-2050-integrated-monitoring-and-reporting-program

Bauman, D., Fortunel, C., Delhaye, G., Malhi, Y., Cernusak, L., Patrick Bentley, L., & Rifai, S. W. (2022). Tropical tree mortality has increased with rising atmospheric water stress. *Nature*, *608*, 528–533.

Bertacchini, E., Liuzza, C., Meskell, L., & Saccone, D. (2016). The politicization of UNESCO World Heritage decision making. *Public Choice*, *167*(1–2), 95–129. https://doi.org/10.1007/s11127-016-0332-9

Burdon, P., Drew, G., Stubbs, M., Webster, A., & Barber, M. (2015). Decolonising Indigenous water 'rights' in Australia: Flow, difference, and the limits of law. *Settler Colonial Studies*, *5*(4), 334–349. doi: 10.1080/2201473X.2014.1000907

Chen, C., MacLeod, J., & Neimanis, A. (Eds.). (2013). *Thinking with Water*. Montreal and Kingston: McGill-Queen's University Press.

Cohen, J. J. (Ed.) (2013). *Prismatic Ecology: Ecotheory beyond Green*. Minneapolis: University of Minnesota Press.

Cohen, J. J., & Foote, S. (2021). Introduction: Climate change/changing climates. In J. J. Cohen & S. Foote (Eds.), *The Cambridge Companion to Environmental Humanities* (pp. 1–10).

Cook, M. (2023). *A River with a City Problem: A History of Brisbane Floods*. Brisbane: University of Queensland Press.

Dale, A., George, M., Hill, R., & Fraser, D. (2016). *Traditional Owners and Sea Country in the Southern Great Barrier Reef – Which Way Forward?* Report to the National Environmental Science Programme. Reef and Rainforest Research Centre Limited, Cairns (p. 50). Retrieved from https://www.cairnsinstitute.jcu.edu.au/traditional-owners-and-sea-country-in-the-southern-great-barrier-reef-which-way-forward/

DeLoughrey, E. (2019). *Allegories of the Anthropocene*. Durham: Duke University Press.

Dillon, R. (2002). Seeing the sea: Science, change and Indigenous sea rights. *Maritime Studies*, *2002*(123), 12–16.

Dobrin, S. I. (2021). *Blue Ecocriticism and the Oceanic Imperative*. New York and Abingdon: Routledge.

Frank, S. (2023). *A Poetic History of the Oceans*. Leiden: Brill.

Giblett, R. (2018). *Environmental Humanities and Theologies: Ecoculture, Literature and the Bible*. Abingdon and New York: Routledge.

Harper, J. (2017). *The Dry*. Pan Macmillan.
Hau'ofa, E. (2008). *We Are the Ocean: Selected Works*. University of Hawaii Press.
Heiss, A. (2022). *Bila Yarrudhanggalangdhuray*. Sydney: Simon & Schuster.
Hills, J., O'Kane, D., & Boswell, R. (2022). Blue heritage in the Blue Pacific. In R. Boswell, D. O'Kane, & J. Hills (Eds.), *The Palgrave Handbook of Blue Heritage* (pp. 273–302). Cham: Palgrave Macmillan.
Hofman, K., & Hughes, K. (2018). Protecting the Great Barrier Reef: Analysing the impact of a conservation documentary and post-viewing strategies on long-term conservation behaviour. *Environmental Education Research, 24*(4), 521–536. https://doi.org/10.1080/13504622.2017.1303820
Hughes, T. P., Day, J. C., & Brodie, J. (2015). Securing the future of the Great Barrier Reef. *Nature Climate Change*, *5*(6), 508–511.
Iovino, S., & Oppermann, S. (Eds.). (2014). *Material Ecocriticism*. Indiana University Press.
Jones, R. (2017). *Slow Catastrophes: Living with Drought in Australia*. Clayton: Monash University Press.
Jue, M. (2020). *Wild Blue Media: Thinking through Seawater*. Duke University Press.
Langton, M. (2006). Earth, wind, fire and water: The social and spiritual construction of water in Aboriginal societies. In B. David, B. Barker, & I. J. McNiven (Eds.), *The Social Archaeology of Australian Indigenous Societies*. Canberra: Aboriginal Studies Press.
Lankester, A. J., Bohensky, E., & Newlands, M. (2015). Media representations of risk: The reporting of dredge spoil disposal in the Great Barrier Reef Marine Park at Abbot Point. *Marine Policy*, *60*, 149–161. https://doi.org/10.1016/j.marpol.2015.06.009
Lloyd, R., White, P., & Brennan, C. (2020). Escaping water: Living against floods in Townsville, North Queensland, from Settlement to 2019. In S. McKinnon & M. Cook (Eds.), *Disasters in Australia and New Zealand: Historical Approaches to Understanding Catastrophe* (pp. 99–117). Palgrave Macmillan.
MacNally, R., Nerenberg, S., Thomson, J. R., Lada, H., & Clarke, R. H. (2014). Do frogs bounce, and if so, by how much? Responses to the 'Big Wet' following the 'Big Dry' in south-eastern Australia. *Global Ecology and Biogeography*, *23*(2), 223–234.
McCallum, C., Walker, R. S., & Dawes, W. R. (2010). Impacts of climate change on groundwater in Australia: A sensitivity analysis of recharge. *Hydrogeology Journal*, *18*(7), 1625–1638.
Mentz, S. (2015). *Shipwreck Modernity: Ecologies of Globalisation, 1550–1719*. Minneapolis: University of Minnesota Press.
Mentz, S. (2020). *Ocean*. London: Bloomsbury.
Mentz, S. (2021). Ice/water/vapor. In J. J. Cohen & S. Foote (Eds.), *The Cambridge Companion to Environmental Humanities* (pp. 185–198). Cambridge: Cambridge University Press.
Mentz, S. (2023). *An Introduction to the Blue Humanities*. New York and Abingdon: Routledge.
Morton, T. (2010). *The Ecological Thought*. Cambridge: Harvard University Press.
Neimanis, A. (2017). *Bodies of Water: Posthuman Feminist Phenomenology*. London: Bloomsbury.
Neumeier, B. (2020). Introduction. In B. Neumeier & H. Tiffin (Eds.), *Ecocritical Concerns and the Australian Continent* (pp. 1–14). Lanham and London: The Rowman & Littlefield Publishing Group.
Oppermann, S. (2019). Storied seas and living metaphors in the blue humanities. *Configurations*, *27*(4), 443–461.
Oppermann, S. (2023). *Blue Humanities: Storied Waterscapes in the Anthropocene*. Cambridge: Cambridge University Press.
Quigley, K. (2023). *Reading Under Wreckage: An Encrusting Ocean*. London: Bloomsbury.
The Reef 2050 Long-Term Sustainability Plan. Retrieved from https://www.dcceew.gov.au/parks-heritage/great-barrier-reef/protecting/reef-2050-plan
Roulière, C., & Egerer, C. (Eds.). (2022). *Water Lore: Practice, Place and Poetics*. Routledge.

Shek-Noble, L. (2021). 'An Indigenous sovereignty of the imagination': Reenvisioning the great Australian novel in Alexis Wright's *Carpentaria*. *Genre*, *54*(2), 195–219.
Smith, J. L., & Mentz, S. (2020). Learning an inclusive blue humanities: Oceania and academia through the lens of cinema. *Humanities*, *9*(3). doi: 10.3390/h9030067
Tarte, D., & Hughes, T. (2020). *Review of State Party Report on the state of conservation of the Great Barrier Reef World Heritage Area*. Report prepared for the Australian Marine Conservation Society. Retrieved from https://independent.academia.edu/DiTarte
Taylor, L. (1990). The Rainbow Serpent as visual metaphor in Western Arnhem Land. *Oceania*, *60*, 329–344.
Te Punga Somerville, A. (2017). Where oceans come from. *Comparative Literature*, *69*(1), 25–31.
Turner, M. G., Calder, W. J., Cumming, G. S., Hughes, T. P., Jentsch, A., LaDeau, S. L., & Lenton, T. M. (2020). Climate change, ecosystems, and abrupt change: Science priorities. *Philosophical Transactions of the Royal Society B*, *375*, 1–11.
Wallace-Wells, D. (2019). *The Uninhabitable Earth: Life After Warming*. Penguin Random House.
Walpole, L. C., & Hadwen, W. L. (2022). Extreme events, loss, and grief – an evaluation of the evolving management of climate change threats on the Great Barrier Reef. *Ecology and Society*, *27*(1).
Webber, B., & Martins, F. (2018). *Big Fella Rain*. Broome: Magabala Books.
Whitehouse, H., Watkin Lui, F., Sellwood, J., Barrett, M. J., & Chigeza, P. (2014). Sea Country: Navigating Indigenous and colonial ontologies in Australian environmental education. *Environmental Education Research*, *20*(1), 56–69.
Wright, A. (2006). *Carpentaria*. Artarmon: Giramondo.

PART 1

Australian identities through the blue

1

BLUE COUNTRY

Nurturing meaningful relationships in discontinuous environments

Vincent Backhaus, Nalisa Neuendorf, Lokes Brooksbank, and Tahnee Innes

The importance of *place* for each of the co-contributors is varied and contested. We acknowledge Yawuru-nation (Broome in Western Australia),[1] Kalkadoon and Kiwai places across Daru and Parama Islands, Erub, Murray Islands in the Eastern Torres Straits (Mt Isa, Australia) and Papua New Guinea, Gimuy, Meanjin (Cairns, Brisbane in Australia) and Manus Island, Papua New Guinea, Port Moresby, as well as the Elema People of Orokolo and Toaripi language groups in the West and East Gulf Province of Papua New Guinea. For this reason, *place* is fluid, turbulent, hot, and cold and continues to be, not only for us four co-authors, but more broadly for a People of Knowledge wherever they are dependent on work, family, history, story, and legacies of settler colonial violence. As such, *place* for social relationships is and always will be a tricky and strategic *endeavour*[2] for the co-authors of this chapter. Like waves, we think about charting and navigating ways forward, around and through the impact of our relationships and the impact of specific words that help us describe relationships. Our understanding of relationships cycles through multiple crests and troughs of multiple waves evolving over time yet rolling back through deep history as an awareness and understanding of everyday experiences. What we privilege in this claim is our connection to and with the notion of being a People of Knowledge. The complexity of defining a People of Knowledge is borne from experience across the Pacific and Oceanic regions, continental (Australia), and the in-between of island landscapes we share as connected co-authors to place and places, people, and communities. The notion of a People of Knowledge responds to the collective of our ways of thinking through and with our research, our standpoints, and our positionality within being family, community, academic, and professional members who live on and across Country. We navigate a continuity of knowing the legacies of discontinuity that still hold impact in our everyday.

DOI: 10.4324/9781003365501-3

The entry point into a chapter discussion on blue spaces is developed from these foundations of forming, undoing, and maintaining good relationships. We see this through an upwelling and downwelling of currents that place us in positions of negotiating our relational connections to epistemes that sit within, outside, and in between institutional and community spaces. The four authors wish to share a perspective from our previous respective doctoral fieldwork sites as we reflect as post-doctoral academics and professionals and raise how the forming of good social relationships with things shapes a narrative not only for our research and work, but in the ways our narratives can speak to the broader emergent qualities found in the storylines of a People of Knowledge.

Blue Country itself is grounded in the holistic interpretation of relationships between land, sea, and sky and the people, places, and things that foster the integrity of a collective understanding of blue spaces (Hau'ofa, 1994; Johansson-Fua, 2022; Langton, 2006). In other words, Blue Country is water and wherever it exists as a relational quality. Epeli Hau'ofa suggested 'we are the ocean' (Hau'ofa, 2008): this context illustrates the basis for how water, which is us, and our understanding of relatedness govern our ways of knowing and being within particular social forms associated to and embedded within place and belonging. Being the ocean means what we would call our *Blue identity* or the *Countried identity* of the places we connect with through our experiences as a People of Knowledge. A *Blue* or *Countried identity* responds and adapts to change and can never be defined by static representations of being. We are active in placemaking, expressing an agency to understanding and resisting connections and relationships to historical legacies taught, people, places, and things. We are seeking alternatives and challenging normative assumptions placed upon our being.

In our chapter, the ways water becomes the metaphor but also the literal connections water plays in our bodies, in the air, on the land, and across the sea reveal the emergent layer and lens for understanding relationships, social or otherwise. We wish to contribute a perspective to what Kanaka Maoli scholar Karin Amimoto Ingersoll demonstrates is an awareness of our oceanic literacy connected to place through a becoming of seascape epistemology (Ingersoll, 2016). We wish to illustrate how we work through watery perceptions to explore the qualities and characteristics of relationships to people, place, and the personhood of things in the understanding of a People of Knowledge existing with Blue Country. We wish to share our scholarly and personable identities within Blue Country, which represents us and our relationships to all things across time.

Towards Blue Country relationships

The four of us acknowledge a diversity of places, mobilities, and experiences that have nurtured our individual scholarship as researchers and professionals. Indeed, we are part of a generation of emerging post-doctoral researchers like those before us who are answering with their lives, the questions that give our lives meaning beyond what we are taught by the settler societies through which we live and move (Meyer, 2013).

Through an ontology of the everyday, relational impacts help us to learn, reflect, or unlearn and then relearn ways that support our ability to exist within the tension of living in institutional, organisational, regional, and remote spaces (Backhaus, in press). We are mindful also of shadowing institutional networks just as they shadowed our ancestors through, for example, the exploration of the Pacific in an earlier period:

> [I]mperialism thrived on the networks of information, power and influence that spanned the globe in the nineteenth century. Imperial careers, webs of missionary control, and extending lines of political influence connected the peripheries of empire to the metropole, underpinning and buttressing formal channels.
>
> *(Mar, 2015, p. 342)*

Blue Country is reflective of this relational ether, whereby thinking through how to navigate and negotiate the existence of our histories, emotions, families, and lives is constantly nurtured through continuous and discontinuous ways that interact in the everyday to form distinct patterns of meaning. At times, the surface can be calm despite oceanic currents shifting the historical claims and lives connected to deep time history, the impact of changes in the more recent era, and the legacies of these connections drawn towards an upwelling of relational affect (Diaz, 2016; Eseli, 1998; Hau'ofa, 1994; Mar, 2016; Mar & Edmonds, 2010). At other times, the surface is filled with white cap waves, a surging, moving windswept reaction to the everyday relations of the contemporary and the challenges to negotiating and articulating a voice that honours ancestral legacies that can be heard within the reverberating sounds of waves, wind, and motion (Mertens et al., 2013).

For a People of Knowledge, when we speak of and with Blue Country, we are mindful of the concepts that have and continue to label and represent our thinking and ways of being. Further, we pause, reflexively engaging in the entanglements of debate surrounding our existence that either affirms or denies the everyday existence of a People of Knowledge (see Nakata, 1997; Williamson, 1997). When we speak of continuities, we think about how we as co-authors support the maintenance of ways of knowing and being that are important while also navigating the impact of systemic and institutional processes borne from a history of imperial and colonial expansion. When we talk about discontinuities, this charts the impact of a continuing presence of colonial expansion through the systems and institutional norms that give it agency and force us to see our relationships to people, place, and things as objectively verified and scientifically alienable. This presence is maintained in our lives and consciously ruptures and challenges the continuity of our everyday – the way we wish to think about our inalienable relationships between ourselves and the world around us. What we achieve from this push and pull is the capacity to innovate and navigate this tension as a form of agential responses of resistance and adaptation within the social sphere of our lives.

Manulani Aluli Meyer suggests 'the world is not black and white, we just *think* it is' (Meyer, 2013, p. 250). We, as a People of Knowledge researchers, confront this complexity and importantly make tentative and cautious steps towards developing relationships with the conceptualisation of the blue and how it is formed. We ought not to simply accept and reproduce epistemic legacies like new wine in old wine-skins, but to think differently about the kind of Story we narrate. As a People of Knowledge, we know this kind of Story has been done before in our region as part of a continuing geo-political agenda impacting upon our ways of knowing and being (Battiste, 2000; Hau'ofa, 1994; Mar, 2015, 2016; Mar & Edmonds, 2010; Moreton-Robinson et al., 2021; Tuhiwai-Smith, 1999). Mar's (2016) work on decoloniality across the Pacific region as a historically *blue space* highlighted the way European people and knowledges and Pacific Peoples' understanding of the sea and oceanic territories privileged certain boundaries, both marine and economic, to ensure benefits flowed to the select few who could exploit the region, its peoples, resources, technologies, and knowledges. This resulted in Pasifika/Pacific Peoples of the region undoubtedly not figuring as majority benefactors of terrestrial and marine wealth exploitations created at the time. This raises important questions for us as authors tasked with a heavy but small contribution into the context of the diversity, complexities, peoples, knowledges, and histories encapsulated by blue spaces. We cannot speak on behalf of all People of Blue Country, nor would we wish to. By suggesting we all belong to a People of Knowledge, we share existence with our neighbours and the way they may also contribute understanding of Blue Country in a holistic, nutritive, and meaningful way to express a kin connection across blue spaces of which we acknowledge our Guesthood among neighbouring Pacific People.

Pasin and Luksave within Blue Country

Pasin and Luksave are linguistically derived words from Tok Pisin (tok meaning talk; pisin meaning a creolisation of English). Tok Pisin is a language which has become lingua franca within Papua New Guinean social realities for communication but equally extends across Melanesian and Pacific spaces as a creolisation to convey social meanings. Previous work by the co-authors of this chapter developed a Melanesian narrative of Pasin and Luksave and the way social relationships have been evaluated by people and community as part of maintaining those good social relationships in the everyday (Backhaus et al., 2020). Pasin can be thought of as a set of principles guiding an assessment of actions interpreted by people. The verb term Luksave itself forms an acknowledgement of actions by people as they observe each other. Someone showing or demonstrating good Pasin (*gudpla pasin*) and/or Luksave can be evaluated by the community as having *good* morality. However, this notion also finds salience across Pacific and First Nation regions in different forms. Contiguously, the use of creole languages – a successful response to altered circumstances borne from discontinuous experiences – has shaped ways of acknowledging and evaluating good relationships (Nakata, 2023). Yumpla Tok, a creole emerging from Torres Strait Islander peoples' engagement with English, for example, also adopts the term Pasin as Ilan Pasin, a way of showing and understanding good relational intentions (Shnukal, 1988, 2002). Solomon Island creole speaks of *Wansolwara* – one salt water which can refer to connectedness (Johansson-Fua, 2022). This connectedness illustrates deep and long-held connections between landscapes, islandscapes, seascapes, and airscapes. These scapes hold the stories traversing large distances, connecting people and places to the understanding of the everyday of social realities.

What is revealed by this understanding are the empowered ways important values regarding good relationships can continue to be observed and followed despite the legacies of settler colonial history. It is through this lens that we embed Pasin and Luksave into Blue Country by highlighting how as co-authors we acknowledge the importance of good relationships to people, places, and things as we build understanding of seeing good relationships as Blue Country and placing our sense of belonging within blue identities.

The stories we share in this chapter speak of an underlying current of mobilities and movements across land, sea, and air scapes to understand and work with social relationships as they form and unform through an ontology of the everyday of our personal and work life, and previous scholarship. Uncle Peter Eseli understood the importance of speaking back to the corpus and challenging the status quo and the definitions of Islander imaginaries. He surmised the importance of speaking about intentional things – *Ilan Pasin things* – by raising points of departure from the corpus and making visible 'issues of kinship and landscape, seascape and skyscape – the relationship of clan to territory – which are fundamental to the traditional Islander worldview' (Eseli, 1998, p. 1). These kinds of deep connections guide us as co-authors to navigate the discontinuities that continue to shape our blue identities in Blue Country.

Four narratives of Blue Country

The following stories are developed through a methodological framework of Storylines (Backhaus, 2018). Storylines are a distinct Traditional Knowledge approach to think about how values are shared through meaning. Building on the notion of Storywork and narrative inquiry (Archibald, 2008; Backhaus et al., 2020; Wyld & Fredericks, 2015), Story not only conveys phenomena but social values for a good life; Story can also enact good meaning or intention (Peschard, 2007). Story is also named, to give proper meaning for the relationship between Story and Country. Respecting this relationship for Story guides connections to place, person, and thing. The adoption of the word Storylines reflects how rivers and streams on Country are shaping country through channels that guide water flows across continental Australia and out across the continental shelf below the current coastal waterline (Backhaus, 2018). The littoral and riparian boundaries of these channels shape relationships, and carry life, water, peoples, and things – connecting places and stories as an epistemic and ontological flow of connectivity (Langton, 2006). Water becomes a Storyline that illustrates the connections between river flows that span out across the continent and the sea, becoming a conduit for meanings to connect with deep oceanic currents. These meanings return over Country as rain to replenish river flows once again. Storylines as such are temporal and cyclical, and the following four stories emerge with this awareness. Each of us as co-authors reflects on our previous doctoral fieldwork situated in various locations of the Pacific and Australian region. The Storylines move across a region, as each of us connects from our institutional places to Story places where community reside and back again. We consider geographical areas including far northern Queensland and Papua New Guinea, connected by the Torres Strait Islands, and the Solomon Island neighbours as markers of Blue identities of a Blue Country. We begin in Papua New Guinea.

Highways of water connecting to place

The ocean and rivers are a 'highway' to the people of the Western portion of the Gulf Province of Papua New Guinea (PNG) given limited infrastructure and remoteness of place. They are integral to connecting related peoples, scattered across a significant land mass that is the Gulf of Papua. I went to Baimuru, an ex-colonial outpost or Government station approximately 338 kilometres from the National capital, Port Moresby, once described to me by a Local Government official as 'isolated from all things. Cut off by river system(s)'. It is a place where I have paternal links, where my father's first cousins were born and raised, some 100 kilometres west of our ancestral village. What surprised me most about my travel to get to my destination by dinghy and outboard motor was how at home I felt being on the water, first the ocean and then on the mighty Pié River on which Baimuru is situated. As the biggest port outside of the Provincial Capital of Kerema, Baimuru is considered 'the station' for the Western Gulf peoples, and as one Uncle stated, 'it is our station, so we come and go'. The relational links to place through my father's people helped me to navigate a new experience, yet our connectivity to each other through the physical – the land and water spaces – developed that relationality beyond familial links. Upon my arrival to Baimuru I was already known because of my family. Situating me within the context of my family links meant identifying me as an 'Orokolo' woman – from a specific place as part of a particular clan or people group. Having that connection and being identified as such, even though I've never actually lived in the village, ensured people then situated how they began to relate to me during my fieldwork: as a daughter, sister, extended family member, or tambu (in-law or affine). It made the necessary links to people that I would otherwise have to develop on my own, were I an 'outsider'. Embedded in place as a woman of the Gulf positioned me as an 'insider' as much as an 'outsider'. With that comes a sociality that requires my own social action (luksave) and ways of doing and being (pasin) that frame my interactions with others, and theirs in return. Having been identified as a woman from this place (the Gulf Province in the broader sense) identified me in a way that enabled a certain relationality to have existed prior, before I even started developing my own individual links with others.

Homeplace at the river

Over two days in 2018, artists, elders, and other young people from Girringun Aboriginal Art Centre went down to the Murray River in Queensland to make mija on the bank of the Murray River. The Arts Manager explained that the purpose of building mija by the group was to facilitate cultural transmission, and to photograph a panoramic background featuring three mijas for Girringun's exhibition at the Cairns Indigenous Art Fair that year. The Dyirbal word 'mija' was described in Dixon (2017) as a noun with 'a wider range of meaning than any other in the language' (p. 192). For example, mija is simultaneously intangible and tangible: it refers both to one's 'homeplace' and the material 'hut' that one has built. I have previously argued that mija building on the Murray River for Girringun people was not just an intangible process, but also the stamping of a material object upon the landscape – in the backyard as it were of the European Australian landowner. Yet today, a mija as 'hut' would be viewed by most as a temporary art piece, thus it is ultimately less threatening to the white landowner. As Jirrbal elder Aunty Emily said of the three mijas that were eventually built, 'we can come visit them if we want. But they'll just stay there like that. Maybe until the river come and wash it away'[3]. Of interest for this chapter is that Aunty Emily, and other Jirrbal and Girramay women, had been camping for several weeks near the mija site. They were settled at the top of the Murray Riverbank, in their traditional country, with a few camp beds, fireplace, and lean-to shelter. This river camp thus reflected the broader meaning of mija as also 'homeplace' – one's traditional country. The camp could only be accessed through private property so, demonstrating the women's strong historic and diplomatic relationship, access for them had been allowed by several white farming families through private property.[4] When I asked one of the Jirrbal women how this arrangement with the farmers had come about, she explained: 'if you're good to them, they're good to you'. The Tully and Murray Rivers are critically important for Jirrbal and Girramay people living in Murray Upper, North Queensland. As Jirrbal elder Ernie Grant once said, 'we always come back to the river'. Mija building, then, was just another instance of 'coming back'. Indeed, when Girramay senior elder Uncle Claude Beeron first greeted me on his Country near the Murray River Falls, he did not say to me, 'Hello, how are you?' as you would in English. Rather, he asked in Dyirbal, 'Where is your mija?'. He thus used the term 'mija' to enquire as to my Country and identity (before asking my name). In this way, 'family on the river' (as I had also described river activity) was not so much temporary or liminal as it was central. An excerpt of a conversation between three senior Girramay and Jirrbal elders further describes this centrality, as well as the concerted effort it took to keep the Murray River 'central' to one's mija – one's homeplace and ever-present traditional country:

TB:	*We still got it in us, but the land is taken away. And scrub and river. It's there but we gotta go to it.*
EM:	*Go ask people, 'Can we get past through there?,' then get down the river.*
CB:	*That's getting to be a pain that one.*
EM:	*I don't care who's there. I just drive a car right around the headland, just to go to the river. Nobody can stop me from the river.*
TB:	*Why would we want to go in the . . . knock all the sugar cane, bananas down . . . we just going through their road just to the river.*
CB:	*Why you walk through there? Because you used to walk through there when you was a kid. And you parents took you through.*
TB:	*Famous track.*
CB:	*Famous track we still follow them track, same way you know.*

Relationality and mountain-scapes

Research conducted for my doctoral thesis took place in the petroleum-impacted regions of the Hela and Southern Highlands Provinces of PNG. While the use of Pasin and Luksave was not explicitly intended as a methodological tool in my research, it emerged as an important layer that supported the storying and data collection process, particularly at the community level, in project-impacted areas. It enabled a smoother, more meaningful, and respectful connection to research participants. This connection was primarily facilitated through intergenerational links between my father and community members, who then subsequently recognised myself as an extension of this prior relationship. The relationships that my father had fostered in his time working in the region enabled myself as a researcher to scaffold new connections and fluently build rapport. On multiple occasions, while in the field, this was recognised through comments from community members such as, 'Ah yu son blo John, em gutpela pasin olsem yu kam behainim em lo kisim ol stori lo mipela long halpim mipela lo viles' – 'Ah you're John's son, you are showing good pasin by following in his footsteps to collect our stories to help us here in the village'.

As I continued my research, I realised that the process of relationality embedded within the good pasin I was showing supported my endeavours as a PNG researcher to foster a unique method of interaction. The good reputation and relationships created by my father working extensively in the region created a recognition that enabled social interaction, which otherwise would have taken some time to establish. Additionally, in industry contexts, this acknowledgement and recognition played a significant role in securing stories and gaining access to corporate and expert research participants. Although the notion of Pasin and Luksave was still emerging as a research method at the time, it contributed significantly to enhancing the ethics, efficiency, and accessibility of the data collection process. Pasin and Luksave were enacted at multiple stages of my research process, facilitating ad hoc conversations or discussions with corporate, community, and expert participants, providing reasonable access to corporate and industry information in real time from the field, and being welcomed into homes and communities based on intergenerational connections that formed lasting relationships still maintained today. The key learning for myself through this experience was a deeper understanding of the continuity of relationality supported by pasin and luksave – while I was not my father, in these contexts and interactions I embodied the respect and relationships which he had formed over his time there.

The sounds of the sea

From the five songs we sang during music practice, most times we started singing with Au Emeret Ghe and then followed with choosing from Pe Dali, Ad Asor, Lovah Ad, and Baba waia. Each gave a rendition of the landscape and seascape of experience held by the Eastern Torres Strait Islands towards the landing of the London Missionary Society (LMS) and subsequent missionisation of the Straits. Emeret Ghe gave me the biggest realisation of the LMS – which fostered my intimate connection to the song. Emeret Ghe spoke of the landing, the placing of the foot on the sand when the boat was beached by the missionaries. Emeret Ghe, while conveying the story of the landing, gave me a tonal sense of moving across the waves, rising and lowering in polyphonic accentuates like a boat in the water charting a path to arrive on an island from distant shores, and the placing of meanings like a foot, and how this story is enculturated into Meriam Mer language. The story begins with the choir master attempting to adjust the tone and articulation of the lyrics. We began: 'Iesu ra mir . . .' ['No wait, it needs to be higher in pitch, we start up here'] said our choir master. ['Stop, here let me show you – Iesu ra mir . . . Like that, okay? Try again']. We began again. ['Start again'] he says as the guitar was strummed. 'Iesu ra mir – ged ge aumeida' – we imagined in these first lines of the song the missionary arriving, lifting his leg up from inside the boat, and stepping over the gunwale. 'Ged ge bazauda' – the missionary's foot coming down outside the boat and landing, touching, placing upon the sand. 'Egared Torres Strait' – the word had arrived, and a sound like no other was heard across the island-scape.

In those first moments of learning this song I was slow and quieter, and the other boys louder and tuned into singing the song before. I hoped my voice was blending into their sounds as I held my pitch, hoping for a commensurate level – to be in tune!

I held the sound more so than the phonetic shape of the words. Kiwai owera (Kiwai thinking – Kiwai speak) filled my head and mouth as I held onto the sound shapes of my mother tongue, so familiar to me as I was taught from birth, as I reached out towards the sonorant pitch of Meriam Mer, the language of the song

Emeret Ghe. Sometimes it felt like both languages Kiwai and Meriam Mer were the same, but most times Kiwai began in my thinking and Meriam Mer ended within the sound I made from my mouth. In that first time of testament – testament being something I recall as important when giving integrity and moral purpose to why you sing the song – I simply remembered running out of breath before the end of the verse and having to take another breath, tripping up and losing place in the words and mucking up the timing and rhythm. My musical cadence with Meriam Mer struggled at either being too late or too early with breathing, singing, or even thinking through the words and the meanings being expressed. I felt the challenge of shaping the sounds or rather shaping myself to relate within the oral cavity of my mouth – the sound of relationship. In other words, the sonoracy of my efforts began to manage more effectively the phonetic intention and, rather than choking myself in the larynx at the start as I struggled with the pitch, I relaxed and felt the sounds move deeper into the chest and belly to my diaphragm, towards the place of what I called magumo muro – Kiwai words for deep knowledge. I did manage after the first choir practice to sense more and more how to connect with the sounds and lyrics as each practice session began. I could begin to feel this sense of being tingling across my chest as the frequency of vibrations spread out from this central trunk of sound around my neck and throat, reaching into my diaphragm and reverberating outwards – I felt how the vibrations languaged my body through an energisingly alternate sonorant form, helping me to think and connect through song to the islands and the Stories and people told through song, breath, feeling, and imaginaries of island waterscapes.

Conclusion

Giving life and voice to Blue Country is part of a legacy of scholarship a People of Knowledge continue to develop as a call to our communities to continue to challenge prevailing notions not only about blue spaces, but the way blue spaces can be conceptualised to benefit the few. In many cases, continental Australia and the Pacific region are still seen by the global north as a far-off southern group of islands in the sea. We are not, we are a *Sea of Islands* – highly diverse, complex, and building scholarship that seeks alternatives to a changing world (Hau'ofa, 2008). The places where we each did our doctoral fieldwork remind us to keep maintaining the importance of good relationships. By writing when we cannot regularly and physically reconnect to the communities that supported us and the people that saw us and helped us to get to where each of us is today, we continue to honour their presence in our lives. This connection challenges the ways research often forgets where ideas emerge – from people and from relationships outside institutional frames of reference. The opportunity to write with our experiences, to express and remember these connections, provides a platform to connect broadly with communities in our blue regions and highlights a continuing legacy developed by a People of Knowledge. Furthermore, the sharing of how each of us conceptualised our relationships within blue spaces, opens the opportunity to draw on scholarship in respectful ways, to contribute and share in the research space, the kind of good social relationship building Pasin and Luksave continue to introduce, and moreover speak to the broader act and intention of good relationship building.

Acknowledgements

The co-authors would like to acknowledge the Elders, families, and communities that supported us to develop our thinking, our lives, and our scholarship across time. Without your love, guidance, and advice we would not be able to give back in the way that each of us continues to do in our working lives.

Vincent would also like to acknowledge the Charlie Perkins Scholarship Trust.

Notes

1 Bracketing acknowledges that for Aboriginal and Torres Strait Island people, land was never ceded and naming words are contested.
2 HMS *Endeavour* was a British Royal Navy vessel that Lieutenant James Cook commanded to Tahiti, New Zealand, and Australia, which sailed to our shores in the Pacific transforming how a People of Knowledge came to know, experience, and remember such legacies. Hence, words have very different meanings which we remember and navigate as we share our intellectual scholarship.
3 C. Beeron, T. Beeron, and E. Murray interview, 2015.
4 The camping place ultimately belonged to the family of a young white woman who had gone to primary school with the elders' children.

References

Archibald, J.-A. (2008). *Indigenous Storywork: Educating the Heart, Mind, Body, and Spirit*. UBC Press.

Backhaus, V. S. (2018). Situating the countried existence of critical Indigenous pedagogies & Aboriginal and Torres Strait Islander student's ways of learning [unpublished thesis, University of Cambridge]. https://doi.org/10.17863/CAM.35719

Backhaus, V. (in press). Navigating the everyday of risk and vulnerability: Being an Indigenous early career researcher. In M. Trudgett, S. Page, R. Povey, & M. L. Locke (Eds.), *Indigenous Early Career Researchers in Australian Universities: Our Stories*. Springer Nature.

Backhaus, V., Neuendorf, N., & Brooksbank, L. (2020). Storying toward Pasin and Luksave: Permeable relationships between Papua New Guineans as researchers and participants. *International Journal of Qualitative Methods, 19*. https://doi.org/10.1177/1609406920957182

Battiste, M. A. (2000). *Reclaiming Indigenous Voice and Vision*. University of British Columbia Press.

Diaz, V. M. (2016). In the wake of Mata'pang's canoe: The cultural and political possibilities of Indigenous discursive flourish. In A. Moreton-Robinson (Ed.), *Critical Indigenous Studies: Engagements in First World Locations* (pp. 119–137). University of Arizona Press.

Dixon, R. M. W. (2017). *Dyirbal thesaurus and dictionary across ten dialects, Part one – Nominals*. Language and Culture Research Centre, James Cook University. Unpublished.

Eseli, P. (1998). *Eseli's Notebook (translated from Kala Lagaw Ya into English)* (A. Shnukal, R. Mitchell, & Y. Nagata, Eds.; Vol. 3). Aboriginal and Torres Strait Islander Studies Unit, University of Queensland. http://trove.nla.gov.au/version/46645179

Hau'ofa, E. (1994). Our sea of islands. *The Contemporary Pacific*, *6*(1), 148–161.

Hau'ofa, E. (2008). *We Are the Ocean*. University of Hawai'i Press.

Ingersoll, K. A. (2016). *Waves of Knowing: A Seascape Epistemology*. Duke University Press.

Johansson-Fua, S. (2022). Wansolwara: Sustainable development, education, and regional collaboration in Oceania. *Comparative Education Review*, *66*(3), 465–483. Scopus. https://doi.org/10.1086/720383

Langton, M. (2006). Earth, wind, fire, and water: The social and spiritual construction of water in Aboriginal societies. In B. David, B. Barker, & I. J. McNiven (Eds.), *The Social Archaeology of Australian Indigenous Societies*. Aboriginal Studies Press.

Mar, T. B. (2015). Shadowing imperial networks: Indigenous mobility and Australia's Pacific past. *Australian Historical Studies*, *46*(3), 340–355. https://doi.org/10.1080/1031461X.2015.1076012

Mar, T. B. (2016). *Decolonisation and the Pacific: Indigenous Globalisation and the Ends of Empire*. Cambridge University Press.

Mar, T. B., & Edmonds, P. (2010). *Making Settler Colonial Space: Perspectives on Race, Place, and Identity*. Springer.

Mertens, D. M., Cram, F., & Chilisa, B. (2013). *Indigenous Pathways into Social Research: Voices of a New Generation*. Left Coast Press.

Meyer, M. A. (2013). The context within: My journey into research. In D. M. Mertens, C. Fiona, & B. Chilisa (Eds.), *Indigenous Pathways into Social Research: Voices of a New Generation* (pp. 249–260). Left Coast Press.

Moreton-Robinson, A., Hokowhitu, B., Smith, L. T., Andersen, C., & Larkin, S. (2021). *Routledge Handbook of Critical Indigenous Studies*. Routledge.

Nakata, M. (1997). Who's reading 'misplaced hopes'? *International Journal of Qualitative Studies in Education*, *10*(4), 425–431. https://doi.org/10.1080/095183997237007

Nakata, M. (2023). Indigenous languages & education: Do we have the right agenda? *The Australian Educational Researcher*. https://doi.org/10.1007/s13384-023-00620-0

Peschard, I. (2007). The value(s) of a story: Theories, models and cognitive values. *Principia*, *11*(2), 151–169.

Shnukal, A. (1988). *Broken: An Introduction to the Creole Language of Torres Strait*. Department of Linguistics, Research School of Pacific Studies, Australian National University.

Shnukal, A. (2002). Some language-related observations for teachers in Torres Strait and Cape York Peninsula schools. *The Australian Journal of Indigenous Education*, *30*(1), 8–24. https://doi.org/10.1017/S132601110000168X

Tuhiwai-Smith, L. (1999). *Decolonizing Methodologies: Research and Indigenous Peoples*. Zed Books.

Williamson, A. (1997). Misplaced hopes or misplaced meanings? A rejoinder to Martin Nakata. *International Journal of Qualitative Studies in Education*, 10(4), 433–439. https://doi.org/10.1080/095183997237016

Wyld, F., & Fredericks, B. L. (2015). Earth song as Storywork: Reclaiming Indigenous knowledges. *Journal of Australian Indigenous Issues*, *18*(2), 2–12.

2

POSSESSING AND PROTECTING THE SOUTHERN OCEAN

Connection and mediation in the Antarctic work of Douglas Mawson and Alan Villiers

Alessandro Antonello

In the winter of 2021, the people of Adelaide were disthalassised – oceanically disoriented – by their city's main newspaper. The *Sunday Mail* reported on an old name being newly affixed by the US-based National Geographic Society to its maps: they would now be labelling the expanse of ocean south of 60° south latitude as the Southern Ocean. For Australians, this decision was a curious and vexing one, for they have named the waters lapping their southern shores the Southern Ocean for generations. Instead, if the National Geographic Society was to be believed, Australia's southern coastline to the west of Hobart was now washed by the Indian Ocean. The *Sunday Mail* wondered if that changed definition had 'left us adrift' (Davies, 2021).

In the Australian imagination – and even the broader global imagination – the Southern Ocean's geographies and boundaries are unstable, always being made and unmade. In late 1841, half a decade after the invasion and settlement of South Australia, a committee studying the local whaling industry noted the colony's proximity to the 'Great Southern Ocean', the home of the whales (Hart et al., 1842). A century and a half later, the artist–author Barbara Hanrahan in *The Scent of Eucalyptus* – her 1973 semi-autobiographical novel of childhood life in late-1940s and 1950s Adelaide – wrote of looking from the city outwards to 'the blue that is the Indian Ocean, edged with the names of seaside towns that sound of England and poetry: Semaphore and Tennyson, Grange and Henley, Glenelg and Brighton, Seacliff and Marine Rocks' (Hanrahan, 1973, p. 80).

In the decades that Hanrahan was growing up and later writing of her youth, scientists were increasingly using the Southern Ocean in a specific way. In the middle third of the twentieth century, oceanographers and marine biologists understood that the cold and tempestuous ocean flowing around the whole Antarctic continent was a distinct body of water and habitat. This they named the Southern

DOI: 10.4324/9781003365501-4

This chapter has been made available under a CC-BY-NC-ND 4.0 license.

Ocean (Antonello, 2018). Bounded to its north by the Antarctic Convergence (known more recently as the Polar Front) separating the warmer waters to the north from its colder waters, the Southern Ocean becomes covered in sea ice during winter and plays an important role in ensuring the Antarctic continent's thermal isolation. The Southern Ocean is one of solid as well as liquid water; the ice, in the form of sea ice or icebergs, is both stationary and mobile, at once massive and impenetrable as well as ephemeral and fragile (Riffenburgh, 2007, pp. 234–239, 741–744, 934–956).

Whatever its name or precise boundaries, part of this ocean has been a space of Australian imagination and exertions since the British invasion and colonisation of the continent in 1788 (Antonello, 2018; McCann, 2018). As Tom Griffiths has put it, in the nineteenth century, 'Encountering and surviving these wild westerlies on the Southern Ocean was a rite of passage for all newcomers to Australia, a dimension of their new identity' (Griffiths, 2007, p. 34). From the nineteenth century to the present, Australian scientists, government officials, entrepreneurs, and workers have gone into the ocean with ideas of geopolitically, spatially, and culturally placing Australia at the heart of a larger southern and oceanic region. It is their ideas and experiences which many Australians have drawn upon to vicariously encounter the Southern Ocean. This is a terraqueous history, since the Southern Ocean is not simply a watery space, but a region of land and sea, and the 'atmospheres, vapours, airs, and waters' constantly connecting them (Bashford, 2017, p. 255). It is terraqueous because this is an ocean being claimed as territory too (Bashford, 2017, p. 261).

To explicate the connected logics of possession and protection that have characterised the Australian relationship with the Southern Ocean, this chapter concentrates on two figures covering the early decades of the twentieth century. It looks to the geologist and explorer Douglas Mawson and his expansionist and technological rhetoric and oceanic boundary work from the 1910s to 1930s. It looks also to the writer and one-time whaler Alan Villiers, whose journalistic account of a Norwegian whaling voyage in 1923–4 – and the presence of a small group of Australians on it – enthused the Australian reading public.

Through them we can explore oceanic connections, boundaries, and media. Connectivity is a central theme of oceanic histories, especially in terms of connections outside state and national borders (Bashford, 2021, pp. 923–924). For Australia, the Southern Ocean is a space of connections framed and inflected by the (emergent) nation and state. For a nation without land borders, oceanic spaces and boundaries are among the sites where Australians meet others – neighbours and strangers, both welcome and unwelcome, whether through the boundary making with New Zealand, France, or Indonesia prescribed by international law, or through those 'working' the seas, whether fishing or people smuggling. Thus, the Southern Ocean is not merely connected with Australia in a sense of proximity, but connects Australians to the world, indeed to the planet, since the ocean, as Charity Edwards figures it, is 'vast, mobile, interconnected with all places, systems and processes of this Earth'

(Edwards, 2019, p. 313). Moreover, these connections are not merely spatial or social but temporal. Telecommunications – first, wireless radio telegraphy from ships and stations used both to maintain basic contact as well as to disseminate news and generate publicity and, from the 1970s, satellite communications and remote sensing to collect vast troves of geophysical data – create a space of simultaneity, an ocean continuous with the Australian continent and nation, an ocean in which ships, weather, animals, and sea ice can be tracked and which allows relations in (near) real time.

The themes of media and mediation also flow through connections. The Southern Ocean, like all oceans, is thoroughly mediated and mediatised. Those perceptual chains are not simply in human media but 'milieu-specific', as specific environments mediate sense and meaning, as Melody Jue argues (2020, p. 3). If Antarctic and Southern Ocean histories have important, even defining, elements of isolation, the modern Australian nation's relationship with the Southern Ocean has been connected by telecommunication technologies. This mediation cultivates both proximity and distance. There is 'a wide range of visual, conceptual, political and legal representations', suggests Edwards, but these keep the Southern Ocean 'at-a-distance: encountered mainly through highly curated reportage or controlled science communications', which are 'romantic imperialist, techno-utopian and depoliticised myths that do little to share the diverse ecologies, sensations and imaginaries within this vast territory' (Edwards, 2019, p. 313). But as Rohan Howitt clearly articulates, Antarctica, even before its thorough exploration, was consistently present in late-nineteenth-century Australian newspapers, made close and proximate in the press, testifying to a rich world of Australian ideas about the Southern Ocean and Antarctica (Howitt, 2019). Both Mawson and Villiers understood the power of the press, and their communication of their sensations and feelings within the oceanic milieu through the press via wireless radio was central to Australia's place in the Southern Ocean.

* * *

In the first half of the twentieth century, Australians turned to the geologist and explorer Douglas Mawson to be their guide and interpreter of the Southern Ocean and Antarctic. English-born, Australian-raised, and University of Sydney-trained, Mawson's fame grew after his first journey to Antarctica as part of Ernest Shackleton's 1907–9 British Antarctic Expedition, on which he became one of the first party to find the location of the south magnetic pole. Soon after he returned from that first Antarctic visit, Mawson began to organise an Australian expedition under his own leadership – that expedition would depart in 1911 as the Australasian Antarctic Expedition (AAE). In his justifications for a national Australian expedition, Mawson articulated a vision of the south – a frontier region of ocean, land, and ice – for white settler Australians to occupy, know, and exploit (Hains, 2002). That vision remained remarkably consistent across his decades of scientific work and advocacy.

When Mawson began to articulate his plans for his own Antarctic and Oceanic expeditionary work, the Australian public had a range of ideas and traditions of public discussion to play with (Howitt, 2019; Kawaja, 2010). Ferdinand von Mueller, among the most prominent scientists in Victorian and Australian colonial society in the late nineteenth century, saw the Antarctic and Southern Ocean as a field of Australian exploratory efforts. His vision, particularly articulated in the mid-1880s, saw Antarctica within a larger geographical and regional frame, associated with the exploration of the Australian continent, New Guinea, and Pacific islands. He was a proponent of economic botany and other useful sciences and hoped that scientific travels in southern latitudes would garner useful animals and minerals for Australians to exploit. Mueller's interventions and dreams inaugurated a more serious public discussion about sending an Australian expedition to Antarctica, although this dream did not see fruition until Mawson's 1911 journey (Home et al., 1992, pp. 392–395).

At the turn of the twentieth century, J. W. Gregory, the short-lived professor of geology and mineralogy at the University of Melbourne (1900–4), also offered a substantive vision of Australia in the Southern Ocean. In a major lecture to the Australasian Association for the Advancement of Science in 1904, Gregory spoke of the Australian climate as being controlled by the Southern Ocean – a terraqueous hemisphere. With words and maps, he described the current state of knowledge – and more significantly, ignorance – of the ocean, the intimate links between the Australian continent and the sea, and the climatic and weather cycles shared by ocean and continent. For Gregory, the ties binding Australia and ocean stretched across several time units, from days, to seasons, to years. Anyone listening to or reading Gregory's lecture could not have doubted either the geophysical importance of the Southern Ocean for Australia or the idea that it should be Australians at the vanguard of inhabiting it (Gregory, 1904).

Mawson was sympathetic to these visions of useful science, resource rationalism, and climate connections, deploying them regularly in his Antarctic boosterism. The Southern Ocean in Mawson's and his supporters' vision at this time was at once national and imperial. In the early twentieth century, Australians considered themselves both Britons and Australians. This perpetuated expansionist and imperialist Australian ideas about Antarctica and the south that had developed during the nineteenth century (Howitt, 2019). One of the plans of Mawson's AAE was to write and print a newspaper during their winter confinement at their Antarctic base – a publication that came to be named *The Adelie Blizzard*. It was an outlet for the expeditioners to mediate their own thoughts and experiences and planned as a tool of publicity upon their return north. In May 1913, to mark Empire Day – a jingoistic celebration of the British Empire on Queen Victoria's birthday – one of the men, likely Archie McLean, reflected on their place in empire: 'The twentieth century Empire has a gigantic body . . . her heart has diffused into the multitude of clear-eyed stubborn pioneers who have plunged into the mystery of the "wild" and built up the Colonial Dominions'. Expeditions were one way of preserving and advancing the Empire, 'British tradition', and 'manly sentiment'; the British Empire 'vitalises the world with the purest ideals!' (McLean et al., 2010, p. 29).

A crucial potential benefit to Australia and the broader British Empire lay in exploitable resources. Mawson was always conscious that his explorations and scientific research should lead to useful knowledge (Shaughnessy & Pharaoh, 2021). As a geologist, he was part of a long scientific tradition of seeking knowledge of the Earth for the benefit of humanity. Part of his justification for the AAE was that the Southern Ocean 'abounds in whale and seal life, and if we can establish convenient harbours or discover sub-Antarctic islands there is sure to follow commercial enterprise on these lines' (Mawson, 1911, p. 617). He spent the winter of 1913 writing articles on Antarctica's 'commercial resources' for *The Adelie Blizzard*. Four long articles between May and October described whales, seals, and penguins – an entirely orthodox account of whales and seals and the products humans made from them, but nevertheless central to his, and the larger public's, imagining of the Southern Ocean. From early on, though, Mawson was sensitive to over-exploitation, as were many others. Of sealing, he called attention to the early nineteenth-century 'rape' of South Georgia's fur seals (McLean et al., 2010, p. 122). In later years, he would point to the depredations on animal and bird populations at Macquarie Island, noting that 'the hand of man has alas! cast a shadow over [the Island's] myriad inhabitants, and wrought irreparable havoc'. He argued that Australia should possess these territories, for it was only Australia that would 'ensure the continuance of species' (Mawson, 1922, p. 40). For Mawson, wildlife protection and wise-use conservation would be achieved through Australian possession.

In a time of rapid technological change, Mawson also saw the place of new technologies in allowing an Australian presence in the Southern Ocean. The emergent technology of radio was crucial for him (Hains, 2002, pp. 49–52; for the larger global context, see Wenzlhuemer, 2016, pp. 171–173). While ideas of Australian–Antarctic proximity had an older valence, for Mawson, the fact that the region was 'within wireless telegraphic distance of our borders' meant that 'this region has a special call upon Australasians' (Mawson, 1911, p. 610). Wireless radio underpinned something like a simultaneity for the whole area. It was not simply that news of heroic deeds could be beamed across the ocean for the entertainment or edification of Australians, but that the radio would allow a real expression of the climatic unity of the southern hemisphere, as prefigured by J. W. Gregory. 'Wireless weather stations', Mawson thought, would be important for 'Australasian weather reports' (Mawson, 1911, p. 617). The first edition of *The Adelie Blizzard* in April 1913 carried news of the first wireless message from Antarctica – a great feat in their eyes, achieved in spite of the 'terrific blizzards' they lived with. Mawson and the other expedition members thought that the wireless would allow rapid dissemination of new scientific knowledge – and indeed, receipt of new scientific knowledge while expeditioning (McLean et al., 2010, pp. 16–17).

Returning to Australia in late 1913, Mawson's expedition was hailed for its geographical and scientific achievements, as well as his heroic feats and survival. The expedition was defined by its tragedy, since Mawson's two sledging mates, Xavier

Mertz and Belgrave Ninnis, died on their journey – Mawson's survival seemed miraculous. Within a year, though, the nature of heroism and geopolitics would change forever, with the beginning of war in Europe. Attention would turn away from the Southern Ocean and Antarctica towards the front lines.

After the war's end in 1918, attention began to turn once more towards the south. Mawson resumed his place as a leading interpreter of the frozen region – touring, lecturing, and writing constantly – although he was not the only man that Australia turned to for expertise on the Southern Ocean in the interwar years. In the whaling industry and whalers' labours, we find a less remembered but important moment in Australia's inhabitation of the Southern Ocean. In November 1923, the mothership of the Ross Sea Whaling Company, *Sir James Clark Ross*, sailed up the Derwent River to dock at Hobart. The company was led by the famed Norwegian whale magnate Carl Anton Larsen, who had pioneered the industry at South Georgia from 1904, which had inaugurated a half century of slaughter that drove whale populations to the brink of extinction. Larsen was hoping to profitably expand his operations to Antarctica's Ross Sea, to the south of New Zealand (Basberg, 2007). The arrival of the *Sir James Clark Ross* in Hobart was greeted by the local paper, *The Mercury*, as a near world-historical event. The newspaper greeted the ship as ranking alongside 'world-famous vessels' like *Fram* and *Terra Nova*, which had carried the Norwegian Roald Amundsen and the Briton Robert Scott on their race to be first to the south pole (Anon., 1923).

In port it met a Hobart public enthusiastic for oceanic life. The city had been a hub of whaling in the early nineteenth century, in the decades after the invasion and colonisation of lutruwita (Tasmania). From it, whalers sailed to hunt near Australian shores or came into port for provisioning during longer hunting journeys. In the 1920s, Tasmanian parliamentarians and aldermen wanted business for their small state, working men wanted work (some wanted a bit of adventure too, to go with it), and a reading public was eager for stories. The docks bustled with onlookers, hungry for connection with a world-straddling ship destined for the South's ice-choked seas.

Alan Villiers was one young man who launched himself upon the Southern Ocean by joining the ship's company. Born in Melbourne in 1903, he had been enchanted by the city's port and docks from his youth: at 11 years of age, he tried to join one of the square-rigged barques docking there. While in later life he spun myth and mystery around his youth, he probably joined his first ship in January 1920, not yet aged 17. After a few years sailing the world ocean, he fell from high on a ship in Port Lincoln, South Australia. Following his serious injury, he eventually found himself in Hobart at the start of 1923 and with a job at the city's newspaper, *The Mercury* (Lance, 2009, chapters 1–2).

When Villiers saw *Sir James Clark Ross* and its five attendant whale catchers, *Star I* to *Star V*, sail up the Derwent River, he signed on at once as a whaler's labourer. At first, he had little thought of writing about his experiences on the whaling ship in Antarctica, chasing only the seagoing experience in treacherous and barely plied waters. But he convinced his employer, although he was neither a journalist nor an author, to publish some of his reports. He cabled news of the Ross Sea Whaling Expedition's movements and successes throughout the voyage.

Towards the end of the whaling expedition, Villiers' report from New Zealand, published on 5 April 1924, caught the public's attention. His story, first published in *The Mercury* and then syndicated more widely, led the newspaper to commission more articles (Villiers, 1924b). Between 10 and 28 May 1924, he published a fifteen-article series on his voyage from Hobart to the Ross Sea via Macquarie Island and returning to New Zealand via the Campbell Islands. The series was syndicated on the mainland too, where the Sydney *Sun*, Adelaide *Register*, Melbourne *Herald*, and Brisbane *Courier* ran the series for a hungry public. So popular was the series that a book of the articles was produced in Hobart, which Villiers later revised and enlarged for an American publisher (Villiers, 1924h, 1931).

Between December 1923 and February 1924, short bursts of words were wirelessly despatched from the Southern Ocean: the sightings, or lack of sightings, of whales; encounters with ice; and the struggles of navigation. Nearing the end of the voyage, Villiers telegraphed: 'Probably this was the most extensive, the fastest, and the most daring voyage ever made in this region' (Villiers, 1924a). For his Australian audience, Villiers wrote of an ocean with Australians in it, but not within an especially nationalist frame. The eleven Australians on board worked alongside the mighty 'Norsemen', the Norwegians who comprised most of the workers on the world's whaling ships. The Australians were participating in 'the most interesting, most spectacular, and at the same time, the most dangerous of all maritime pursuits' on a ship with 'the most powerful wireless installation afloat in any merchant vessel' (Villiers, 1924c). It was only in the last article in his series that he reflected of the eleven Australians that 'they manfully upheld the honour of their own young Australian flag down there in that terrible waste of ice and gale' (Villiers, 1924g).

In the Ross Sea – 'the coldest and stormiest sea in the world, and the literal end of the world of waters' (Villiers, 1924c) – Villiers was attentive to the atmospheres and colours about him in the 'foamy' seas (Villiers, 1924d). The vista was not always only blue and white; sometimes green and brown water and ice intruded. The ship was 'harassed' by 'heavy ice fogs'; 'clouds of frost-smoke' made the water seem 'afire' (Villiers, 1924e). The bloody thrashing of the harpooned whale turned the ocean into a 'seething cauldron' (Villiers, 1924f) – another medium pervading the terraqueous world, another milieu of sensations.

The Hobart public devoured Villiers' narrative. The descriptions of hard work, challenging nature, wondrous animals, and brutal slaughter excited readers. At a public reception in June 1924, the head of the Tasmanian Marine Board remarked upon the reception of Villiers' despatches:

> In one instance a housewife, who usually in the evening took down the Bible from the shelf and read a chapter, picked up "The Mercury" during the currency of the articles, and read about the Frozen South. (Laughter.) He [the Marine

> Board head] had asked a friend the other day if he happened to be reading them, and the answer was, "I can't get my 'Mercury,' as the maid in the kitchen gets it first, and by the time she has read the Frozen South article I have no opportunity of doing so." (Laughter.)
>
> *(Anon., 1924)*

Through Villiers, Hobartians and Australians had felt Antarctica's 'proximity' (Anon., 1924). His writings allowed for a cultural consumption of the whales and the ocean, just as the whalers were commodifying the ocean and its denizens in more literal, material ways.

Villiers was not the only Australian to join the Norwegian whaling ships in the Ross Sea. Between 1923 and 1931, two Norwegian companies used Hobart as a port to access the Ross Sea whaling fields. The first was Larsen's Ross Sea Whaling Company until 1926, followed by the Polaris Whaling Company with the *N.T. Nielsen-Alonso* – 132 young Australians, mostly Tasmanians, joined the predominantly Norwegian crews to hunt whales in the tempestuous Southern Ocean across eight voyages (Stoddart, 2017). Work on a whaling ship was gruelling and dangerous. The blue humanities must keep the labouring body of the sailor present in its analysis (Blum, 2015, p. 27); the working class have a place in Australia's Southern Ocean history, although very much under-studied (Maddison, 2014).

Only some who followed Villiers chose to write of their work. Before the 1924–5 season the need for whalemen was 'flashed on the screens at all the picture theatres' in Hobart, prompting 'three or four adventurous youths to hazard the trip' (Tilley, 1925a). Athol Tilley, who had joined Villiers on the first voyage, re-joined with plans to make a film of the trip, taking with him 10,000 feet of film. Although that film did not eventuate, he published six articles in *The Mercury* on his return. Tilley explored the prospect that the industry he was participating in might wipe out the whales; Villiers, by contrast, was clearly in awe of the whaling industry, the whales, and their ocean home, and did not especially consider the fate of whale populations. Tilley reported a conversation with Carl Anton Larsen (who died on the voyage):

> I asked him if he did not think the whales would be altogether exterminated if they were assailed on such a huge scale. "There is no chance of that," he replied. "There are thousands upon thousands of big blue whales . . . There are enough whales to keep the whalers going for many years to come."
>
> *(Tilley, 1925b)*

For the attentive reader of *The Mercury*, this reported conversation came soon after reports of a lecture by Sidney Harmer – the recently retired director of the London Natural History Museum – warning that 'unlimited whaling had invariably resulted in the reduction of the number of whales to a vanishing point' (Anon., 1925).

Tilley, like Mawson and Villiers before him, was also journeying in a wirelessly activated ocean. For the readers of *The Mercury*, he explored the limits of both the Ocean to their south and the technological challenges of reaching its full extent. While the *James Clark Ross* was anchored in Discovery Inlet, their wireless, 'splendid set though it was,' as Tilley put it, 'was practically useless . . . The ice seemed to have a tremendous deading effect upon the wireless waves, and it was quite impossible to pick up Awarua (the New Zealand high power station . . .)' (Tilley, 1925c). Yet in other parts of the ocean, the wireless was a powerful tool of placing themselves in time and space. They could hear broadcasts from Melbourne and Sydney, and news broadcast from New Zealand and other stations 'was intercepted and issued on board in bulletin form' (Tilley, 1925d).

* * *

At the end of the 1920s, while some young Tasmanians followed in Villiers' wake to join the Norwegian whalers, Douglas Mawson returned to the Antarctic. In late 1929, after years of lobbying and planning, he set sail on the first of two planned summer cruises for the British Australian New Zealand Antarctic Research Expedition (BANZARE), which he commanded. The expedition's driving purpose was discovering new lands and claiming them for Australia. The possessiveness of earlier years thus moved into a formal register. And with that shift in frame, the fellow workers and neighbours of the Southern Ocean – the proud 'Norsemen' that Villiers wrote of, among others – became competitors.

The drama of competition and boundary-making in the ocean animated the public's appreciation of BANZARE. Its first cruise – aboard *Discovery*, the ship built for Robert Scott's first Antarctic expedition (1901–4) – began in Cape Town in October 1929. Before embarking, Mawson faced a press storm that imagined a race to claim Antarctic territories between him and the Norwegian Hjalmar Riiser-Larsen commanding the *Norvegia*, a sealing vessel refitted for exploration. Newspapers in Australia, Britain, and further afield covered the supposed race. Mawson left port without meeting any Norwegians, but with the expectation that he would encounter them somewhere in the Southern Ocean (Price, 1962, pp. 25–27). Almost three months later, on the evening of 14 January 1930, at about 47° east longitude, *Discovery* met *Norvegia*. Riiser-Larsen had hitherto been a telegraphic spectre. Cables had been arriving from early January advising Mawson of Riiser-Larsen's territorial claims along the Antarctic coast; those cables also advised that the Norwegian government had disavowed the claims (Price, 1962, pp. 60–70, 74). Only the day before, Mawson and a party had finally stepped onto land – an island just offshore from the continent – to take possession of their discoveries with a formal ceremony (Price, 1962, pp. 70–72). When their ships met along the edge of the pack ice, Riiser-Larsen came aboard *Discovery* to talk of his exploratory work. A friendly encounter, they met to partition the ocean and their spheres of interest and influence; *Discovery* would return to the east and *Norvegia* to the west (Mawson, 1930, pp. 550–551). Wireless radio carried reports of the meeting, the boundary making, and the atmospheres of encounter to newspapers around the world.

Both Mawson and Villiers activated the Southern Ocean as a place continuous and simultaneous with the Australian continent and nation. When the Australian government proclaimed the Australian Antarctic Territory in 1933, the Southern Ocean became a space connecting formalised territories of government control. In his presidential address to the Australian and New Zealand Association for the Advancement of Science in 1935, Mawson made it clear that the biological riches of the region ought to be exploited, but that only effective possession, with systems of 'wise control and development', could prevent it being 'raped' and being 'left for ever desolate' (Mawson, 1935, p. 33). For Mawson, Australia, as a core member of the British Empire, was naturally the one to possess and protect. Villiers' and Mawson's words and actions conjured a drama of people, heroes, villains, animals, airs, waters, and sensations that allowed Australians to look south and see an ocean for the taking.

Acknowledgements

This work was supported by an Australian Research Council Discovery Early Career Award (DE190100922) funded by the Australian Government.

References

Anon. (1923, 26 November). Ross Sea whaling expedition. Mother ship arrives at Hobart. Largest floating factory in world. *The Mercury, 3.*

Anon. (1924, 4 June). The frozen south. Marine board meets Mr A. J. Villiers. *The Mercury, 6.*

Anon. (1925, 31 March). Sir Sidney Harmer on need for regulation. *The Mercury, 9.*

Antonello, A. (2018). The Southern Ocean. In D. Armitage, A. Bashford, & S. Sivasundaram (Eds.), *Oceanic Histories* (pp. 296–318). Cambridge University Press.

Basberg, B. L. (2007). Larsen, Carl Anton. In B. Riffenburgh (Ed.), *Encyclopedia of the Antarctic* (pp. 583–585). Routledge.

Bashford, A. (2017). Terraqueous histories. *The Historical Journal, 60*(2), 253–272.

Bashford, A. (2021). World history and the Tasman Sea. *American Historical Review, 126*(3), 922–948.

Blum, H. (2015). Terraqueous planet: The case for oceanic studies. In A. J. Elias & C. Moraru (Eds.), *The Planetary Turn: Relationality and Geoaesthetics in the Twenty-First Century* (pp. 25–36). Northwestern University Press.

Davies, N. (2021, 20 June). Shifting sea has left us adrift. *Sunday Mail* [Adelaide], 68–69.

Edwards, C. (2019). The ocean in (planetary) excess. *Dialogues in Human Geography, 9*(3), 312–315.

Gregory, J. W. (1904). *The Climate of Australasia in Reference to its Control by the Southern Ocean*. Whitcombe and Tombs.

Griffiths, T. (2007). *Slicing the Silence: Voyaging to Antarctica*. UNSW Press.

Hains, B. (2002). *The Ice and the Inland: Mawson, Flynn, and the Myth of the Frontier*. Melbourne University Press.

Hanrahan, B. (1973). *The Scent of Eucalyptus*. Chatto & Windus.

Hart, J., Hagen, J., & Baker, J. (1842, 4 January). Transactions of the Statistical Society. Report on whaling. *Southern Australian* [Adelaide], *3.*

Home, R. W., Maroske, S., Lucas, A. M., & Lucas, P. J. (1992). Why explore Antarctica?: Australian discussions in the 1880s. *Australian Journal of Politics and History, 38*(3), 386–413.

Howitt, R. (2019). *Ideological Origins of the Australian Antarctic, 1839–1933* [PhD thesis, University of Sydney].
Jue, M. (2020). *Wild Blue Media: Thinking through Seawater*. Duke University Press.
Kawaja, M. (2010). *The Politics and Diplomacy of the Australian Antarctic, 1901–1945* [PhD thesis, Australian National University].
Lance, K. (2009). *Alan Villiers: Voyager of the Winds*. UNSW Press.
Maddison, B. (2014). *Class and Colonialism in Antarctic Exploration, 1750–1920*. Pickering & Chatto.
Mawson, D. (1911). The Australasian Antarctic expedition. *The Geographical Journal, 37*(6), 609–617.
Mawson, D. (1922). Macquarie Island and its future. *Papers and Proceedings of the Royal Society of Tasmania*, 40–54.
Mawson, D. (1930). The Antarctic cruise of the 'Discovery', 1929–1930. *Geographical Review, 20*(4), 535–554.
Mawson, D. (1935). Presidential address: The unveiling of Antarctica. *Report of the meeting of the Australian and New Zealand Association for the Advancement of Science, 22*, 1–37.
McCann, J. (2018). *Wild Sea: A History of the Southern Ocean*. NewSouth.
McLean, A., Leane, E., Pharaoh, M., & McEwin, E. (2010). *The Adelie Blizzard: Mawson's Forgotten Newspaper 1913*. The Friends of the State Library of South Australia in association with the Friends of Mawson at the South Australian Museum.
Price, A. G. (1962). *The Winning of Australian Antarctica: Mawson's B.A.N.Z.A.R.E. Voyages 1929–31*. Angus and Robertson.
Riffenburgh, B. (Ed.). (2007). *Encyclopedia of the Antarctic*. Routledge.
Shaughnessy, P. D., & Pharaoh, M. (2021). Mawson's views on use of natural resources in the Antarctic and sub-Antarctic. *Polar Record, 57*, e31.
Stoddart, M. (2017). *Tassie's Whale Boys: Whaling in Antarctic Waters*. Forty South.
Tilley, A. (1925a, 25 April). Adventures of a whaler. *The Mercury, 15*.
Tilley, A. (1925b, 29 April). Adventures of a whaler. *The Mercury, 5*.
Tilley, A. (1925c, 4 May). Adventures of a whaler. *The Mercury, 10*.
Tilley, A. (1925d, 7 May). Adventures of a whaler. *The Mercury, 5*.
Villiers, A. J. (1924a, 25 February). The whaling expedition. *The Mercury, 6*.
Villiers, A. J. (1924b, 5 April). The whaling expedition. *The Mercury, 7*.
Villiers, A. J. (1924c, 10 May). To the frozen South. *The Mercury, 10*.
Villiers, A. J. (1924d, 13 May). To the frozen South. *The Mercury, 9*.
Villiers, A. J. (1924e, 16 May). To the frozen South. *The Mercury, 5*.
Villiers, A. J. (1924f, 17 May). To the frozen South. *The Mercury, 10*.
Villiers, A. J. (1924g, 28 May). To the frozen South. *The Mercury, 5*.
Villiers, A. J. (1924h). *To the Frozen South*. Davies Brothers.
Villiers, A. J. (1931). *Whaling in the Frozen South: Being the Story of the 1923–24 Norwegian Whaling Expedition to the Antarctic* (New edn). Robert M. McBride & Company.
Wenzlhuemer, R. (2016). The ship, the media, and the world: Conceptualizing connections in global history. *Journal of Global History, 11*(2), 163–186.

3

WRITING THE MORE-THAN-HUMAN HISTORY OF NORTHERN AUSTRALIA'S MANY WATERS

Environmental history, the blue humanities, and the challenge of entanglement

Claire Brennan

Australia is the driest inhabited continent, and as a result water is a clear and present theme in its environmental history. The continent is often described in terms of water (generally its absence, sometimes its presence), and its waters of significance are its permanent bounding oceans, its coastal fringes, and its inland waters including creeks, rivers, swamps, lakes, and dams. While some of those inland bodies of water are more often absent than present, they remain significant features that define the continent. The coastal fringe (by definition land entangled with ocean) is home to the vast majority of the Australian population and is an important area of environmental change.[1] The potential to see Australia as a continent defined by water presents an opportunity to extend recent scholarship within environmental history. Choosing to examine the Australian continent in watery terms opens the potential for valuable interactions between environmental history and the blue humanities.

This chapter surveys the presence of water in Australian history, identifies ways in which water's unreliable presence has informed distinct strands of environmental history, and explores the potential for connections between environmental history and the blue humanities. Concentrating on North Queensland, this chapter suggests that exploring the entanglements of continent and coast, and the place of inland waters in the history of Australia, offers new insights into the environmental history of Australia's north. Frames of reference are significant in promoting (and silencing) lines of enquiry, and northern Australia is itself a flexible category, although it engulfs very nearly half the continent and a wide range of landscapes (and waters). The environmental history of northern Australia rewards examination and offers the opportunity to expand the blue humanities' understandings of watery connections and real environments.

DOI: 10.4324/9781003365501-5

The absence of present water in Australian history writing

Much Australian history is written in the cities of Canberra, Melbourne, and Sydney, and as a result it reflects the interests and experiences of the southeast corner of the continent.[2] Those interests and experiences tend to be landlocked as well as geographically restricted. In a work dealing with intellectual exchange, historian Warwick Anderson has noted Australia's 'deeply repressed oceanic memories' and previous, futile attempts to include Pacific consciousness in the telling of Australian history (2015, p. 436). Similarly, Ian Hoskins' recent work placing Australian history in a Pacific context makes the observation that 'Though we live in an island nation, Australians do not generally acknowledge their historical connections to the ocean—at least beyond sand and surf' (2021, p. 3). The fact that the broad survey of Australian history that follows is not immediately familiar to most Australian readers reinforces Hoskins' point that Australian historians rarely recognise Australia's Pacific connections. In contrast to Hoskins' work, Grace Karskens' 2009 history of early Sydney offers an example of the way in which the first Australian settler colony dominates Australian historical consciousness, which focuses on land, not water. Karskens, conversely, defined the settlement on the Cumberland Plain by its waters, noting the region's three harbours (Sydney Harbour, Botany Bay, and Broken Bay), and the significant inland waters of the girdling Hawkesbury-Nepean River, the Parramatta River, and the waters flowing from the Paddington Ridge. In addition, Karskens noted that colonists arrived by water, that they used rivers as ways to penetrate the interior of the continent, that until the 1820s all New South Wales' towns and villages were coastal or located on rivers, and that the original colony 'faced the waters' (Karskens, 2009, p. 2). However, when connecting the early settlement with its hinterlands Karskens considered those hinterlands in purely continental terms. For her they consisted of the Cumberland Plain (now beneath the city), the local rivers, and the Blue Mountains (which bound the Plain). Karskens did not include the ocean on the settlement's doorstep within this analysis. This was despite the significance of that ocean to early Sydney, which was identified by Australian historian John Young in the 1960s, who also recognised the emergence of the colony as a pivotal event in Pacific history (Young, 1967). While Karskens' work helped recover Sydney's deep and significant Aboriginal history, its focus was terrestrial, and it did not discuss the significance of the Pacific that lapped the shore of the settlement, and that connected it to established economies and Pacific resources. Karskens' later history of the Hawkesbury-Nepean River is masterful, but its focus on a river again disengages the colony of New South Wales from the Pacific Ocean that nurtured it (Karskens, 2020).

From outside mainstream Australian history, the significance of the continent's coastal connections is slightly more evident. Tracey Banivanua Mar's work on

Pacific travellers to Sydney acknowledged that the resources of the Pacific were seen as vital when planning the colony there and that Sydney was 'a node in a wider, island-hopping Pacific world of travel and mobility during the first decades of the nineteenth century' (Banivanua Mar, 2015, p. 341). However, Banivanua Mar's work sprang from her interest in the history of Melanesian labour in Queensland, and in the impacts of settler colonialism on Indigenous peoples; it did not draw Pacific Sydney into the mainstream national narrative of Australian history, and it restricted the significance of Sydney's Pacific connections to the early decades of the nineteenth century. It also belongs more to Pacific history than to Australian history, and it acknowledged the significance of historians of New Zealand and the Pacific in identifying 'the mobility and interconnectedness of Pacific peoples' (Banivanua Mar, 2015, p. 342). Among those significant historians is James Belich, and in his monumental history of New Zealand he argued that Sydney owed its very existence to the desirability of New Zealand's timber and flax, coupled with effective Māori resistance against the incursions of European explorers. Belich alerted readers to Sydney's Pacific connections, noting it was a base for European whalers, sealers, and missionaries operating throughout the Pacific and arguing that in the early period of European settlement Australia and New Zealand were not 'markedly separate places' (Belich, 1996, p. 131). For many industries and their workers 'the distinction between Australia and New Zealand was artificial': the Tasman Sea formed a connection rather than a barrier (Belich, 1996, p. 132). Belich mapped out a mobile, maritime world invisible to Karskens, and even to Banivanua Mar. This blue view of Australia has also been developed by historians of the Pacific alert to Australia's maritime entanglements, but histories of Australia's trading forays into the Pacific remain marginal in Australian historiography.[3] The maritime links (both actual and imaginary) forged by shipping routes across the Pacific had also faded from popular memory, until recently exhumed by Victoria Kuttainen and her collaborators (Kuttainen, 2018). Such oceanic connections threaten to overflow the boundaries of national histories, tying national stories to transnational forces and demonstrating the ways in which reconsidering water can cast new light on the past.

The presence of absent water in Australian environmental history writing

While histories of Australia do not dwell on the significance of saltwater to European colonisation of the continent, Australian environmental histories are aware of the quirks of Australia's freshwater. Early in his overview of Australian and Pacific environmental history Don Garden's description of the Australian continent focused on its lack of water:

> The majority of the continent consists of the "typical" Australian landscape—flat and generally with the red sandy soil and stunted vegetation of an arid

> climate. Only a narrow triangle in the southwestern corner of the continent and a sliver running from the south-eastern corner along the eastern coast offer the comfortable mix of a reasonably moist climate, areas of slightly mountainous terrain, and a mixture of forest, woodland, and wooded grasslands.
>
> *(Garden, 2005, p. 2)*

With this description Garden emphasised that the environment most familiar to the vast majority of Australia's population (and its historians) is anomalous in terms of the continent as a whole: people tend to live where the rain falls. This situation had earlier been summarised by environmental scholar George Seddon, with his description of Australia as 'a small country with big distances' (Seddon, quoted in Griffiths, 2019). For Australians clustered in the southeast corner, most of the continent is something to be traversed when travelling between well-watered areas. It is the well-watered regions that are familiar, and it is the well-watered regions that luxuriate in extensive written histories. Histories of the inland, of the Australian 'wilderness', and of maritime industries have been written, but they are few in number when compared to the historical texts devoted to more densely peopled regions (including economically important agricultural and pastoral regions) and to the activities of politicians and urbanites. The views offered by Seddon and Garden demonstrate the ability of environmental history to disrupt this human-centred view and to offer a richer, more-than-human story of the continent, a challenge ably taken up in historian Michael Cathcart's *The Water Dreamers* (2009).

As Cathcart's work illustrates, Australia's lack of water, often conceptualised as drought, caused grief to Australia's European explorers and then to its European settlers. The pain caused by the absence of expected water has found a place in environmental historiography which contains a rich vein of work examining Australia's human history of drought, and its more-than-human history of climate (Anderson, 2014; Garden, 2009; Jones, 2017; Morgan, 2015; O'Gorman, 2021). Such works have moved the study of water and its absence away from an assessment of economic impact and agricultural potential and are part of an evolving conversation about human engagement with Australia's environments and its waters. That conversation includes discussion of excessive as well as absent water. While Dorothea Mackellar's 1908 observation that Australia is a land of 'droughts and flooding rains' risks devolving into banality, it reflects the links between the presence and absence of water across the Australian continent (State Library of New South Wales, n.d.). Australian environmental historians have followed those links: the historiography of floods has expanded recently, particularly with reference to Queensland. Margaret Cook's *A River with a City Problem*, and the responses of historians to the 2019 Townsville (Queensland) flood, show the maturation of Australian environmental history in its examination of Australia's erratic rainfall, as well as of its rivers and its wetlands (Cook, 2019; Lloyd et al., 2020). Inland waters are entangled with oceans and the large weather systems they drive, and as environmental historians work to determine the bounds of their topics, they demonstrate the fluid nature of Australia's capricious waters.

Water and the history of northern Queensland

Northern Australia remains sparsely settled. While approximately 45 per cent of the Australian continent lies above the tropic of Capricorn, only 5 per cent of the Australian population lives here (Bolleter, 2019). Within the region of northern Australia, the coastal strip of Queensland is itself an anomaly. It differs from the rest of northern Australia in supporting significant urban populations and regions of intensive agriculture (in contrast to the extensive pastoralism that occurs away from the coast). The history of this region is rich in topics which might be explored under the aegis of blue humanities, some of which are explicated in this section. Where it has water, it has population, and refocusing accounts of northern Queensland on the entanglements of inland waters, land, and sea offers a new and productive framing for historical studies of the region. The environments of the north, and the interactions between people and environment that occur here, while only partially explored by historians, already add considerably to Australia's environmental history.

The Great Barrier Reef is a globally significant entity that dominates North Queensland's maritime world. While histories of Australia tend to be dominated by terrestrial concerns, the Reef as a region has its own historiography. Iain McCalman has been instrumental in historicising the people involved with the Great Barrier Reef, telling human stories of European exploration of the region (McCalman, 2009, 2016). Rohan Lloyd's work on the Reef shifts the focus of analysis from the activities of individuals to the perceptions and activities of interested groups including scientists, miners, and mariners (Lloyd, 2020, 2022; Lloyd et al., 2017). This ties into a larger historiography concerned with maritime enterprise, and with the Reef as a site of economic activity (Lloyd, 2018; Mullins, 2019).

Recent terrestrial histories of northern Queensland, even away from the Reef, also acknowledge the significance of water to the region. The emptiness of northern Australia has long troubled Australian politicians, and much of the rhetoric about its potential for development relates to water, particularly the prospect of damming northern rivers to promote agricultural development, a theme familiar from the colonial period to the present (White & McGregor, 2016). In their works, both Russell McGregor and Lyndon Megarrity look to inland waters, rather than the ocean, following the gaze of politicians and developers. The history of northern Australia's large dam projects and their underwhelming outcomes, from the Ord to the Burdekin, is central to understanding the ways in which settler Australia struggles to understand the continent and its environments (McGregor, 2016; Megarrity, 2018). The prospect of damming surface waters regularly excites journalists and politicians (for example the Bradfield scheme, first proposed in 1938, regularly features in state election campaigns), but it is the underground water of Australia's Great Artesian Basin that has allowed pastoral development to occur in western Queensland. However, Queensland's subterranean water is yet to have its history written in full. The inclusion of inland and underground water within the environmental history of Australia offers an opportunity to expand the interests of the blue humanities beyond an examination of saltwater topics.

Elements of the environmental history of northern Queensland offer the blue humanities a clear rationale for extending its view beyond the coast. In the north, the relationships between ocean and inland regions are muddled by a range of natural phenomena. Northern Australia's inland waters are voluminous yet unpredictable because of the region's exposure to the monsoon, a phenomenon which brings with it the risk of cyclones developing over the ocean and crossing the coast to the interior of the continent. Cyclones are charismatic weather systems – they are named (Yasi, Larry, and Althea are infamous examples), tension builds as they develop and approach the coast, their paths are always uncertain, and they cause immediate and highly visible damage.[4] In the north, charismatic predators as well as meteorological phenomena can cross apparent 'natural' boundaries to threaten humans. Even in the southeast corner of Queensland bull sharks inhabit the Brisbane River, lurking within Queensland's largest city (Dawson, 2012; Werry et al., 2011). Similarly, and in spite of the species' name, the range of saltwater crocodiles is not limited to the ocean but extends inland along rivers. Such charismatic natural hazards mark out North Queensland as a region all its own, and its narratives of environmental hazard and entanglement offer the blue humanities a space to explore the challenges of coexisting with natural systems and establishing suburbia in the north of the Australian continent (Brennan, 2015; Maglen, 2018).

Indigenous perspectives on northern waters

The connections between land and water, people and environment are central concerns within Australia's Indigenous cultures. Those concerns cannot be summarised or universalised, and anthropologists Ilisapeci Lyons and Marcus Barber note that even within the region of northern Australia there is diversity in 'worldviews, knowledge systems, and assets' as well as in experiences of settler colonialism (Lyons & Barber, 2021, p. 57). Engaging with this diversity is a challenge yet to be widely met by settler Australia, and the diversity and dynamism of Aboriginal traditional understandings and actions largely elude officialdom. Historian Heather Goodall has witheringly critiqued tendencies among the judiciary, government agencies, and members of environmental movements to consider the environmental management knowledge of Traditional Owners to be locked in a pre-colonial past, unable to draw on new understandings or evolve to use new tools. In her account of complex and dynamic Indigenous understandings – including knowledge of massacre sites, incorporation of Great Artesian Basin bore sites in stock routes, and the significance of rivers in the 'land' rights movement of the 1970s – Goodall underscores the importance of including water in Australian history if history is to escape the viewpoint imposed on it by settler colonialism (Goodall, 2008).

Studies of watery topics in North Queensland also need to meet the challenge of meaningful engagement with Aboriginal views of fresh and salt waters. The Rainbow Serpent is an aspect of Aboriginal cosmologies familiar within broader Australian culture. It is known for its association with subterranean waters, the sky, rain, and the ocean. Its creative acts include carving out rivers and water holes (Kolig, 1981). In the Gulf of Carpentaria, the Rainbow Serpent can manifest not only as rainbows but also as cyclones and waterspouts (McNiven, 2004). Anthropologist Deborah Bird Rose's exploration of Aboriginal understandings of water, drawn from studies across Australia, notes the significance of all kinds of bodies of water to the history of Australia: 'Rock holes, soaks, wells, rivers, clay pans, water-holding trees, billabongs, springs and the like form part of the subsistence geography of country and invariably part of the sacred geography as well' (Rose, 2005, p. 39).

In North Queensland the significance of Sea Country and saltwater peoples is recognised within the regular assessments produced by the Great Barrier Reef Marine Park Authority, demonstrating at least some official engagement with Aboriginal knowledge of Country (Brennan, 2021). However, the significance of water extends inland beyond the remit of the Authority, and the actions of the Rainbow Serpent in the Gulf of Carpentaria make clear connections between sea and inland waters. Without simplifying or universalising Indigenous understandings of water in northern Australia, it is still possible to note that waters and their associated spirit beings, resources, and Traditional Owners cannot be neatly categorised as belonging solely to either land or sea. Both environmental history and the blue humanities have their work cut out for them learning to engage with the waters of North Queensland and the knowledge of them held by Traditional Owners.

The blue humanities and 'blue' history

Historian Helen Rozwadowski has argued that environmental history is directly connected to an interest in the world's oceans triggered by the first images of the earth from space (Rozwadowski, 2018). If environmental history is fundamentally a blue enterprise, Australia's environmental historians have proved themselves capable of engaging with the continent's human and more-than-human histories of water. The possibility of drawing that work together and extending it is tempting, and blue humanities is a compelling term that offers environmental historians an opportunity to consider water as an organising principle. When outlining the potential of what Steve Mentz first termed 'blue cultural studies' and later blue humanities, Mentz explicitly included historical studies within his blue turn, acknowledging the significance of the development of regional studies of ocean basins by historians (Mentz, 2009). That development has been outlined by historian Jerry Bentley (Bentley, 1999), while the challenges of identifying boundaries that are meaningful rather than arbitrary (and the risk of replacing the 'myth of continents' with the 'myth of oceans') has been explored by Martin Lewis and Karen Wigen (Lewis & Wigen, 2021; Stolte, 2015). In 1993, Pacific scholar Epeli Hau'ofa, in his visionary work *A New Oceania*, demonstrated the potential of conceptualising the Pacific as a vast ocean world rather than a collection of isolated islands and thinking of the Ocean as a highway rather than a barrier. Building on this emerging thalassology, historians using oceans as tools for framing their investigations have uncovered new ways of understanding the past.

Among the social sciences, it is not only history that can contribute to reassessments of the significance of water to human experiences, and it is not only oceans that are 'blue'. Rozwadowski has urged the inclusion of arts and humanities in the movement known as 'ocean literacy', which emerged from the physical sciences (Rozwadowski, 2020). The United Kingdom's National Maritime Museum's involvement with the idea of ocean citizenship in the twenty-first century makes a strong connection between the oceanic turn, real remains, and archaeology, as does the 2018 statement by the UNESCO chair on the ocean's cultural heritage (Fletcher & Potts, 2007; Brito et al., 2018). Australia's environmental historians have shown that not all water worth studying is blue, and in the eyes of sailors water can also be green (coastal) and brown (navigable rivers and estuaries) (Department of the Navy, 2010). In Australia, the murky and often spectral waters of Australia's creeks, rivers, and lakes are important more-than-human habitats and affect the nature of coastal waters. Those coastal waters are integrally connected with the larger ocean. Blue humanities offers environmental historians a rationale for framing their studies around these connections and a challenge to be fluid in their thinking. In return, history offers a bridge between blue humanities and ocean literacy, and between humanities, social science, and science disciplines.

Conclusion

The rising awareness of the blue humanities holds promise for the study of North Queensland, and the environmental history of North Queensland offers opportunities to the blue humanities. Australian environmental historians have demonstrated their willingness to frame studies in terms of water, and northern Australia offers subjects that extend the blue humanities by including absent and present, inland and ocean waters. In North Queensland that distinction between ocean and inland is physically undercut by rivers as well as by large predators and weather systems that cannot be prevailed upon to confine themselves to saltwater environments. The muddied and compromised waters of creeks, rivers, estuaries, and coasts during the Monsoon allow the blue humanities to take on deeper hues. While the holistic integrity of the water cycle is not unique to Australia, it is well recognised in Australian Indigenous conceptions of the extent of water bodies and their reach into Country, and is a topic worthy of further consideration in the north.

North Queensland offers a place to explore both Australian environmental history and the blue humanities. Framing studies using water – its absence, its sometimes over abundance, and its dangers – can offer new insights into the human past of the region. Recognising the entanglement of the region with its fringing ocean revises settler-colonial and nationalist versions of Australian history, offers new insights into existing settlement patterns, and informs critique of continually re-emerging, unrealised developmental ambitions for the region. The history of northern Australia can benefit from using the frames of reference offered by the blue humanities, while recognising the diverse and entangled bodies of water present in northern Australia can in turn connect the blue humanities to social science disciplines and to complex real environments.

Notes

1 The proportion of Australia's population that lives on the coast is considerably higher than the global average and is expected to reach 92 per cent by 2050 (Yang & Kelly, 2015).
2 This is evident from the location of the institutions that participated in the 2018 report on the state of university history in Australia and New Zealand, which are concentrated in the southeast corner of the continent, and which serve to support historical research and publishing (Crotty & Sendziuk, 2018, p. 66).
3 An early example of such a history is Dorothy Shineberg's examination of Australian involvement in the Pacific sandalwood trade (Shineberg, 1967). Like Young's work (Young, 1967) the text is marginal in Australian historiography.
4 For an overview of Australian cyclones, please see: http://www.bom.gov.au/cyclone/tropical-cyclone-knowledge-centre/history/past-tropical-cyclones/

References

Anderson, D. (2014). *Endurance: Australian Stories of Drought*. CSIRO Publishing.

Anderson, W. (2015). Liberal intellectuals as Pacific supercargo: White Australian masculinity and racial thought on the boarder-lands. *Australian Historical Studies*, *46*(3), 425–439. https://doi.org/10.1080/1031461X.2015.1071417

Banivanua Mar, T. (2015). Shadowing imperial networks: Indigenous mobility and Australia's Pacific past. *Australian Historical Studies*, *46*(3), 340–355. https://doi.org/10.1080/1031461X.2015.1076012

Belich, J. (1996). *Making Peoples: A History of the New Zealanders: From Polynesian Settlement to the End of the Nineteenth Century*. Allen Lane, Penguin Press.

Bentley, J. (1999). Sea and ocean basins as frameworks of historical analysis. *Geographical Review*, *89*(2), 215–224. https://doi.org/10.2307/216087

Bolleter, J. (2019, 25 October). You can't boost Australia's north to 5 million people without a proper plan. *The Conversation*. https://theconversation.com/you-cant-boost-australias-north-to-5-million-people-without-a-proper-plan-125063

Brennan, C. (2015). 'An Africa on your own front door step': The development of an Australian safari. *Journal of Australian Studies*, *39*(3), 396–410. https://doi.org/10.1080/14443058.2015.1052833

Brennan, C. (2021). Protecting the environmental integrity of the Great Barrier Reef, Australia. In B. McNicol (Ed.), *Sustainable Planet: Issues and Solutions for Our Environment's Future, Vol. 1* (pp. 371–380). ABC-Clio.

Brito, C., Carvalho, P., Garcia, A., Vieira, N., Bettencourt, J., & Costa, J. (2018). The UNESCO chair on the ocean's cultural heritage: A brief note on oceanic history, science, and literacy. In A. Roque, D. de Paula, J. Dias, L. Goncalves, L. da Fonseca, M. Rodrigues, R. Vasoncelos Junior, & S. Pereira, *Mares e Litorais: Perspetivas Transdisciplinares, Tomo VII da Rede BRASPOR* (pp. 31–43). UERJ.

Cathcart, M. (2009). *The Water Dreamers: The Remarkable History of Our Dry Continent*. Text Publishing Company.

Cook, M. (2019). *A River with a City Problem: A History of Brisbane Floods*. University of Queensland Press.

Crotty, M., & Sendziuk, P. (2018). *The State of the Discipline: University History in Australia and New Zealand*. Australian Historical Association. https://theaha.org.au/the-state-of-the-discipline/

Dawson, C. (2012). 'Voracious monsters': People, sharks and the Brisbane River. *Queensland History Journal*, *21*(10), 653–665.

Department of the Navy, Washington DC. (2010). *Naval Operations Concept 2010*. Defense Technical Information Center. https://irp.fas.org/doddir/navy/noc2010.pdf

Fletcher, S., & Potts, J. (2007). Ocean citizenship: An emergent geographical concept. *Coastal Management*, *35*(4), 511–524. https://doi.org/10.1080/08920750701525818

Garden, D. (2005). *Australia, New Zealand and the Pacific: An Environmental History*. ABC-Clio.

Garden, D. (2009). *Droughts, Floods and Cyclones: El Niños that Shaped Our Colonial Past*. Australian Scholarly Publishing.

Goodall, H. (2008). Riding the tide: Indigenous knowledge, history and water in a changing Australia. *Environment and History*, *14*(3), 355–384. https://doi.org/10.3197/096734008X333563

Griffiths, T. (2019, 3 December). Professor of everything. *Inside Story*. https://insidestory.org.au/professor-of-everything/

Hau'ofa, E. A. (1993). Beginning. In E. Hau'ofa, E. Waddell, & V. Naidu. *A New Oceania: Rediscovering Our Sea of Islands* (pp. 126–139). School of Social and Economic Development, The University of the South Pacific.

Hoskins, I. (2021). *Australia & the Pacific: A History*. NewSouth.

Jones, R. (2017). *Slow Catastrophes: Living with Drought in Australia*. Monash University Publishing.

Karskens, G. (2009). *The Colony: A History of Early Sydney*. Allen & Unwin.

Karskens, G. (2020). *People of the River: Lost Worlds of Early Australia*. Allen & Unwin.

Kolig, E. (1981). The Rainbow Serpent in the Aboriginal pantheon: A review article. *Oceania*, *51*(4), 312–316. https://doi.org/10.1002/j.1834-4461.1981.tb01473.x

Kuttainen, V., Liebich, S., & Galletly, S. (2018). *The Transported Imagination: Australian Interwar Magazines and the Geographical Imaginaries of Colonial Modernity*. Cambria Press.

Lewis, M., & Wigen, K. (2021). A maritime response to the crisis in area studies. *Geographical Review*, *89*(2), 161–168. https://doi-org /10.1111/j.1931-0846.1999.tb00211.x

Lloyd, R. (2018). Optimism unlimited: Prospects for the pearl-shell, bêche-de-mer and trochus industries on Australia's Great Barrier Reef, 1860–1940. In N. Cushing & J. Frawley (Eds.), *Animals Count: How Population Size Matters in Animal–Human Relations* (pp. 73–84). Routledge. https://doi.org/10.4324/9781351210645-8

Lloyd, R. (2020). An extraordinary barrier: European exploration, shipwrecks and early heritage values on the Great Barrier Reef 1770–1860. *History Australia, 17*(1), 40–58. https://doi.org/10.1080/14490854.2020.1717347
Lloyd, R. (2022). *Saving the Reef: The Human Story Behind One of Australia's Greatest Environmental Treasures*. University of Queensland Press.
Lloyd, R., Newlands, M., & Petray, T. (2017). Coral Battleground? Re-examining the 'Save the Reef' campaign in 1960s Australia. *Environmental Sociology, 3*(1), 54–63. https://doi.org/10.1080/23251042.2016.1259604
Lloyd, R., White, P., & Brennan, C. (2020). Escaping water: Living against floods in Townsville, North Queensland, from settlement to 2019. In S. McKinnon & M. Cook (Eds.), *Disasters in Australia and New Zealand: Historical Approaches to Understanding Catastrophe* (pp. 99–117). Palgrave Macmillan. https://doi.org/10.1007/978-981-15-4382-1
Lyons, I., & Barber, M. (2021). Relatedness and co-existence in water resource assessments: Indigenous water values, rights and interests in the Mitchell Catchment, North Queensland. *Australasian Journal of Water Resources, 25*(1), 57–69. https://doi.org/10.1080/13241583.2021.1921331
Maglen, K. (2018). 'An alligator got Betty': Dangerous animals as historical agents. *Environment and History, 24*(2), 187–207. https://doi.org/10.3197/096734018X15137949591945
McCalman, I. (2009). *Darwin's Armada: How Four Voyages to Australasia Won the Battle for Evolution and Changed the World*. Viking.
McCalman, I. (2016). *The Reef: A Passionate History*. Penguin Random House.
McGregor, R. (2016). *Environment, Race, and Nationhood in Australia: Revisiting the Empty North*. Palgrave Macmillan.
McNiven, I. (2004). Saltwater people: Spiritscapes, maritime rituals and the archaeology of Australian Indigenous seascapes. *World Archaeology, 35*(3), 329–349. https://doi.org/10.1080/0043824042000185757
Megarrity, L. (2018). *Northern Dreams: The Politics of Northern Development in Australia*. Australian Scholarly Publishing.
Mentz, S. (2009). Toward a blue cultural studies: The sea, maritime culture, and early modern English literature. *Literature Compass, 6*(5), 997–1013. https://doi.org/10.1111/j.1741-4113.2009.00655.x
Morgan, R. (2015). *Running Out?: Water in Western Australia*. UWA Publishing.
Mullins, S. (2019). *Octopus Crowd: Maritime History and the Business of Australian Pearling in its Schooner Age*. The University of Alabama Press.
O'Gorman, E. (2021). *Wetlands in a Dry Land*. University of Washington Press.
Rose, D. B. (2005). Indigenous water philosophy in an uncertain land. In L. Botterill & D. Wilhite (Eds.), *From Disaster Response to Risk Management. Advances in Natural and Technological Hazards Research* (pp. 37–50). Springer. https://doi-org/10.1007/1-4020-3124-6_4
Rozwadowski, H. (2018). *Vast Expanses: A History of the Oceans*. Reaktion Books.
Rozwadowski, H. (2020). Ocean literacy and public humanities, *Parks Stewardship Forum, 36*(3), 365–373. https://doi.org/10.5070/P536349841
Shineberg, D. (1967). *They Came for Sandalwood: A Study of the Sandalwood Trade in the South-West Pacific, 1830–1865*. Melbourne University Press.
State Library of New South Wales. (n.d.). *Dorothea Mackellar's My Country*. State Library of New South Wales. https://www.sl.nsw.gov.au/stories/dorothea-mackellars-my-country
Stolte, C. (2015). Map-making in world history – an interview with Kären Wigen. *Itinerario, 39*(2), 203–221. https://doi.org/10.1017/S0165115315000431
Werry, J., Lee, S., Otway, N., Hu, Y., & Sumpton, W. (2011). A multi-faceted approach for quantifying the estuarine–nearshore transition in the life cycle of the bull shark,

Carcharhinus leucas. *Marine and Freshwater Research*, *62*(12), 1421–1431. https://doi.org/10.1071/MF11136

White, P., & McGregor, R. (2016, 13 December). The keys to unlock Northern Australia have already been cut. *The Conversation*. https://theconversation.com/the-keys-to-unlock-northern-australia-have-already-been-cut-69713

Yang, S., & Kelly, S. (2015). The use of coastal reservoirs and SPP strategy to provide sufficient high quality water to coastal communities. *Journal of Geoscience and Environment Protection*, *3*, 80–92. doi: 10.4236/gep.2015.35010

Young, J. (1967). *Australia's Pacific Frontier: Economic and Cultural Expansion into the Pacific, 1795–1885*. Cassell Australia.

4

THE COLOUR OF WATER

Mia-Francesca Jones

i.

In North Queensland, a recent tropical cyclone has brought all that was below the water to the surface. The Bureau of Meteorology has assigned this storm a name – male, of Arabic origin. The practice of naming individual storms is a way of keeping lines of communication clear. It makes me think: a storm is like a message, or a messenger, that brings with it information about the state of things.

Today, in Far North Queensland, the Barron River, or the *Bunawurru*, is still, as is the mood. Palm fronds trip their way down the boat ramp. Ankle deep in the river water, the sting of the midges travels from my toes in the thick sand and up my bare leg. I swing the weight of my body into the boat.

The Barron is a nursing ground of barramundi, grunters, jacks, and bream. The Traditional Owners of this stretch of river are the Djabugay Nation, and the five family groups that belong to it – *Djabuganydji*, *Bulwanydji*, *Nyagali*, *Yirrganydji*, and *Gulunydji*.[1] Over a hundred years ago, white settlers performed baptisms in this hot ruddy water. Once, during a storm, a rogue cow was washed down this river to Redden Island where it still lives now, free range. All the other livestock turned up dead and bloated on beaches, legs in the air. Even fenced animals can be disorganised by nature.

Tropical storms can rip roofs off of houses, tear through glass shop fronts, and disorder forests. A cyclone requires a perfect symphony of wind, rain, and sea – according to the naturalist E. J. Banfield, who wrote about two tropical storms in 1918 after three months of heavy rainfall.[2] He described uprooted palm trees, cracked coconuts, and shaken sea birds. Thousands of maritime birds were killed in the havoc. The only species to maintain their numbers after the storms during 1918 were the cave-dwelling swiftlets, safe in their granite homes. At school, we were

DOI: 10.4324/9781003365501-6

taught to take shelter during a cyclone in rooms with the fewest windows and drag gym mattresses along for protection.

After the storm, Banfield wrote: *nature spares no time in her soothing duties.*[3] The very next day the sun shines on newly arranged clearings, germinating seeds of jungle plants dispersed by birds from which a denser forest grows. The mess has a way of cleaning itself up.

Of the developments brought on by the big wind, Banfield wonders whether they crop up not only in connection with plant and bird life, but with the human life of the island as well. The storm cuts deep furrows, raises ridges, amends the shoreline. Perhaps, he suggests, it changes its inhabitants too, with a subtle shift in temperament.

My brother Hal drops the engine into the river. The clunk of the vessel hitting the water triggers a flash of memory: South of France, 2001. My father pays us in francs to collect walnuts from a freezing pond. Hal falls for it, and emerges fist first, holding a wrinkled tree nut like a planet in his palm.

All his life he's been waterlogged.

As a teenager, instead of going to school he went to the surf, and came home after dark, stinking of salt, with sea ulcers festering on his skinny brown legs. Then we moved north, where the reef breaks the surf. There, he took a gun to the water. Spearfishing brings him closest to the marine life he loves, yet his weapon takes it away. His relationship with the reef is one of both kinship and violence. He has swum alongside eels, sea turtles, and infant whales. He has killed fish the size of his young son. Once, he climbed inside of a croc trap to make the boys in the boat laugh, and ignored my desperate pleas to get out – so at one with the waterways, he thinks he is invincible.

When we were in our early twenties, I tattooed two runes on his inner ankle, one above the other, like an equation:

protection
/
water

My mum has always called him the man who traded a paperclip for a house. When we first moved here, he acquired a rusty tinny – long done-up and sold on. His new boat has a larger engine, wider hull, and a shade canopy.

As we pull away from the boat ramp, I know this might be our last trip together.

The charges laid against Hal have hung over the family for two years and finally, they're coming to a conclusion. On bad days I force myself to Google sentencing. I've read: eight years, twenty years, life. It feels impossible to conceptualise the idea of him locked in a place away from the river and the sea, up at Lotus Glen where the sugar cane grows high and the mosquitos hover like black halos around your skull. The colours on the Tablelands are so bright, in the wet season the rain turns everything an almost fluorescent green: the water finds its way into the land.

The colour of the Barron River is dependent on the weather: an angry storm will stir up the sediment and redden the water. On calm days, it is almost teal. The mineralogist Abraham Gottleb Werner created a nomenclature of colours as a reference to describe objects in nature in the pre-photographic era.[4] The best way to describe the indescribable is to place it next to something else. Each hue is prescribed an animal, a mineral, and a vegetable.

The colour of that French water beneath the walnut tree in my memory was Verditter blue (lenticular ore, a stone of green and cyan); so cold it was almost purple. But the colours of nature are so gradual, so various. Werner knew this too; he showed it when he wrote: *When one colour approaches slightly to another, it is said to incline towards it . . . When it approaches very near to one of the colours, it is said to fall, or pass, into it.*[5]

The Barron is like this: the river falls – or passes – into the ocean. The brackish water (spot of a guinea pig, breast of a hooper, iron flint) passes into a wood brown (hazel nuts, mountain wood) and then falls into a leek and bluish green (egg of thrush) at the mouth of the river. From here, the water flows into the ultramarine blue of the open ocean (upper side of the wings of a small blue heath butterfly).

Many writers have been beguiled by the colour blue. Every scrap of blue, according to Maggie Nelson, might be the fingerprints of God – a tarp flapping in the breeze, spilt ink, pebbles, and fish.[6] Rebecca Solnit writes: *The world is blue at its edges and at its depths.*[7] This is the blue I imagine philosopher William H. Glass was referring to in his meditation of blue, when he wrote that *blue is the colour of spaces* – that is to say, blue is the colour of possibility. [8]

In colour theory, blues and violets have the highest frequency, or the smallest wavelength. Hue, chroma, value, saturation, tones, tints, shades all affect the way we perceive colours. Blue and green might have the greatest emotional range: sadness, envy, distance, loss. Reef fish have an ability to perceive colours invisible to the human eye. Some fish, like the Triggerfish, see the same spectrum of colours as humans, but their colour discrimination is different. Their colour perception has blue-bias, and they see all other colours in relation to blue.[9]

Out on the water, the emotional range of the ocean feels bottomless. Steve Mentz writes: *the ocean is an emotional reservoir*.[10] Our shared history, Hal's and mine, is expansive and it feels like it is sinking. Our memories span northern and southern hemispheres, our father's cruelty, these boat trips, our childhood malaise – which is mauve, like a faded bruise.

ii.

An estuary is a place where the river meets the sea.

To be estuarine is to move like a snake.

To contemplate Hal's demise is to write an impact statement to a judge. How can I make succinct the oceanic depth of our childhood and its repercussions? I have tried to write it many times, but it feels impossible.

Our childhood ripples into our adult lives. We don't talk about it, but we swim through it.

When we were young we looked for omens. As we've gotten older, I've grown out of it, but Hal hasn't. He still calls and asks me to send something out into the universe, as though my goodness (against his badness) has some kind of bearing on, or power over, consequence.

I humour him, but secretly, I send no message to the universe, because I've stopped believing in this kind of thing: crystals, weather warnings. Red sky at night, sailor's delight, red sky in the morning, sailor's warning. The day the cops raided his house, the sky flooded crimson. My mother saw protection, like a cloak, but I saw red, the colour of blood.

It feels childish to look for messages in the sky, but then again, even scientists give storms a name, and with a name comes a character and a whole story. In a letter to Sigmund Freud, the lofty French writer Romain Rolland coined the phrase *oceanic feeling*: a sensation of eternity, of being whole.[11]

A mathematician, of all people, contemplated that perhaps human beings start their lives this way – with a sense of wholeness in the universe.[12] She bases her belief on a 1994 study by psychiatrist Ulrich Gebhard, who found that at ages six to seven, a child still believes everything to have a consciousness. He quotes children's phrases like: *the sun shines because she is kind, the clouds want to make it rain.*[13] At age eight this is limited to mobile things, and at twelve only to animals – though Gebhard believes such child-like animism never fully leaves us, but that adults still feed off these past experiences of unity.

This suggests: our enlightened worldview conceals the magical only with a flimsy layer.

As children, we shared a collective imagination, Hal and I. Born only fifteen months apart, our memories begin at the same time. Me younger, my memory better, him older, his memory obscured – our conscious lives align. A burnt down tree at the back of our bush property was an elephant we took turns riding. Once, I snuck out alone – I wanted to have it to myself. But without Hal there, I saw it for what it really was: a hunk of charcoal and ash. It was only in our shared internal life that the animal felt real. If internal life has a colour, then Glass believed it would be blue.[14] Maybe this is what Serpil Oppermann meant when she described the sea's two-fold condition: *both a physical geographical site, as well as a vast domain of imagination that can never be conclusively charted.*[15]

On the water, the child in Hal is reanimated, and I can see him again at five, at seven, diving for walnuts and coming up blue. At twenty-eight, staring with eyes the colour of sky at the open water and its endless possibilities for our imagined past and future.

The mangroves in the estuary give us shelter, like the swiftlet in its granite cave. Crocodiles stick out of the water with scales the colour of a dried wound. Hal knows the river like the back of his hand. Even when the water levels rise after a storm or flood, he knows to swerve the boat where a tree trunk lingers just beneath the surface, scouts the wreckage of an old trawler, notices croc nests in what seem to me nothing more than a pile of mud and straw.

He has created an imaginary chart of the river with these emotional pinpoints of destruction and play, but the ocean is a different story. When the boat moves through the mouth of the river, it is spat into the sea and the water becomes choppy and white. It is no longer soft and malleable; it hits the hull like cement.

We know the reef lies in the distance like a rare jewel of a thing that makes us feel rich for living here. Despite the swamped house that floods in the wet season, despite the rotting wood of the deck, the charges against his name. It is there like a guardian, like a wish, and to visit it like this is a different kind of homecoming.

I try to find words for a sensibility that is estuarial: multiplicity, flowing, mixing, mingling, when one thing becomes another. A person is fluid – there is damaged child and complicated adult. How to chart this within a letter for the eyes of the law?

iii.

Here in the north, there are two seasons: wet and dry, which is to say, there are two moods.

I am visiting from Victoria, in the south east of Australia where the seasons are stark. Hot, dry summers frame black cows against the blonde grass. Crisp, ombre autumns colour the ground yellow, cold winters bring an abundance of garden citrus, spring rain conjures a green that is wet and bright, like paint that is yet to set.

I'll admit that the heat in the north still makes me mad.

Two Christmases ago, I brought my boyfriend, Tom, home for the first time, and Hal took him out on the boat. Afterwards – about Tom – Hal said, well he's very *Victorian*, and I knew what he meant by that: educated, good. Not infected with this tropical madness that seems to make all the boys up here bad. That is: criminal, fun.

The streets were hit that year with flash flooding. After an argument, I drove my dead grandmother's car from my mother's house to Hal's house and the car filled with water up to my knees. My grandmother's cigarette butts were resurrected as they spilled out of the ashtray and sloshed around, turning the water brown. I cried and let my madness drive the car, swerving chaotically around the wide roundabouts, speeding so fast the cane became one colour: white, like eyes. As quickly as everything flooded, the sun came out and turned the water into steam.

Tom said: this is what it's like to be here with you. One minute, a deluge. The next, it is calm.

This is what it's like to be a part of this family.

Of the elements, Aristotle says they have four qualities: hot and cold, wet and dry. Everything is affected by the interaction between them. In every case, he says, heat and cold determine, conjoin, and change things of the same kind, and things of different kind: moistening, drying, hardening, and softening them.[16] Constant becoming and constant destruction.

Putrescence is the end of all things, he says – the end of all natural objects except those destroyed by violence: you can burn flesh, bone, anything else, but the natural course of their destruction ends in putrefaction. He writes: *the sea putrefies quickly when broken up into parts, but not as a whole, and all other waters likewise.* And the body made of water and determined by its own boundary must be either hard or soft – for it either yields or does not.

Along the journey to the reef, the water turns from brackish to salt, though it is hard to pinpoint the exact place where the water becomes completely oceanic. Even specifying a certain concentration of salt to fresh water would be unclear, as this place would vary depending on the time of the day, the strength of the current.

The idea of the reef as a destination itself is unclear. The reef is not a determined place, but a place made up of many places. Collections of corals and fish and organisms that shine in a multitude of colours, some bright, some fading, some flourishing, some dead.

Instead, Hal picks a seemingly random place to stop, based on intuition. There are no waves on the reef – the boat and water have become one again as the water laps softly against the metal shell. The anchor enters the water like our bodies, with a heavy weight. Our flesh is protected by thick wetsuit skin, we use snorkels to breathe and flippers to propel ourselves through the water. It is illegal to hunt fish with breathing apparatus aside from a snorkel: there has to be some even playing field.

Hal swims backwards, the tip of his spear pointed away from me, like a compass or a weathervane, guiding him somewhere I'll never go. He always offers me a gun, but I prefer to be a submerged voyeur: complicit, yet not active in the hunt.

Hal dives down deep and emerges from the water with a cray. As we breach the surface, he throws it into the boat. The crustacean's body is hard on the outside, soft on the inside. When it hits the floor of the boat, it does not yield, but scatters like rice.

The ocean is broken up into parts: coral reef, breeding ground. Divided by lines: okay to fish, illegal to spear. The water is zoned by colour: general use (light blue), habitat protection (dark blue), limited spearfishing permitted in a conservation park (yellow). But things aren't so black and white; these territories leak into one another. Without this knowledge of subterranean worlds or underwater life, it seems like one block colour. All water is messy then, as are our memories.

Hal was a sensitive child. He cried when a pet fish died. Each one had a whole life of its own. As an adult, he has participated in this violence, on the water and the land. When he was charged with it, the whole family became complicit. My mother buried guns over the property line of her rainforest home, but a big rain came, unearthed the burial, and washed them back.

Aristotle explains that bodies which contain water are either foreign or connatural. Foreign, he says, like the water in wool. Connatural, like that in milk. I want to tell the judge this: it's not so clear. It's not so easy to zone a human being like this, to categorise or define them as one thing. But if you wade through the debris, if you wait for it to be washed down river, there is a pearl of truth somewhere. And perhaps it is both things, that we are good and we are also bad, but it is the properties of our mixed-up experience that make us so.

About the cray, Hal says, *that one's just a baby*. He pulls himself into the boat, picks it up, and throws it back. Somehow, that cray has crossed the boundary of two worlds and managed to return home. Bits of splintered shell that cracked in the fall cover the bottom of the boat like crushed glass.

We spend the next few hours hunting for reef fish to smoke later in the ground, which we will serve with sweet chilli sauce. The wings, Hal says, are the best, the cheek, the meatiest.

When the sky starts to turn grey and the wind becomes warm, we know it is time to head back. From my view, there are no boundaries on the open ocean, just blue, blue, blue. Somehow, Hal knows his way back.

We coil through the estuary like smoke, to return in some way to our primordial home. If my brother is the fish that grew legs, found a paperclip, traded it for a tinny, swapped it for an illegal substance and got caught, then our childhood is the wave he rode in on.

Timothy Morton says that nature appears like a ghost at the never-arriving end of an infinite series: *crabs, waves, lightning, rabbits, silicon.*[17] Things blur into one other, brackish becomes salt which dissolves and dries up on our skin, which leaves fingerprints, and traces of our DNA on everything it touches.

What happens when a storm hits is that nothing can be hidden. A storm mixes it all up, spilling like brush water onto a palette, combining all the colours into a clumsy swamp. A storm messes things up but it also leaves a clearing on which to start again and regrow. Except we've learnt from these waterways that nothing is ever clear. There are always particles of debris, sediment, algae, salt. Even a cleared spot of land is rich in soil, seed, worm, history. We can dig it up, but we can't rebury it. All we can do is try to make something new out of it and hope the water will wash us clean.

The colour of that water depends on the mood of the day, the weather, the current, the trail of a boat's engine, the ripple of human legs and arms that thrash and part the water. And colours, like a storm, come with a message. Skin that turns purple is cold. A wound that goes red is infected. Blue lips, red cheeks, dark eyes all say something about the state of things. The reef is like this too, every colour of the coral, of a fish's body, carries with it information or a message of warning. And every act of violence has a consequence. Glass wrote: *every colour is a completed presence in the world*, and when these colours fade, or change, something goes with it.

When we get back to land, Hal pulls the boat onto the truck and winds it up the ramp. I sit in the back while he drives it home, past the Holloways Beach bar, the cement block beach toilets, the palms lining the shore, to the house swamped by the weather with its rotting deck. When we get home, Hal wonders aloud who will feed his fish while he's gone, and the moment is yellow, like the streak in the eye of the kingfisher, or the glare of sun after a storm.

Notes

1 Bunawurru is a contested area of Country. Information for this piece was provided by Djabugay man and Traditional Owner Levi Newberry, via Djabugay Council.
2 Banfield, E. J., *Last Leaves from Dunk Island* (n.d.). Retrieved 30 November 2022 from https://gutenberg.net.au/ebooks03/0300451h.html
3 Banfield, E. J., *Last Leaves from Dunk Island*. (n.d.). Retrieved 30 November 2022 from https://gutenberg.net.au/ebooks03/0300451h.html
4 *Werner's Nomenclature of Colours as adapted by Patrick Syme to Zoology, Botany, Chemistry, Mineralogy, Anatomy, and the Arts* (1814). Second Ed. Art Meets Science, Read and Co. Books.
5 *Werner's Nomenclature of Colours as adapted by Patrick Syme to Zoology, Botany, Chemistry, Mineralogy, Anatomy, and the Arts* (1814). Second Ed. Art Meets Science, Read and Co. Books.
6 Nelson, M. (2009). *Bluets*. Wave Books.
7 Solnit, R. (2005). *A Field Guide to Getting Lost*. Penguin.
8 Glass, W. H. (1991). *On Being Blue*. David R. Godine.
9 Champ, C. M., Vorobyev, M., & Marshall, N. J. (2016). Colour thresholds in a coral reef fish. *Royal Society Open Science*, *3*(9). doi: 10.1098/rsos.160399
10 Mentz, S. (2019). Shakespeare and the blue humanities. *SEL Studies in English Literature 1500–1900*, *59*(2), 383–392. https://doi.org/10.1353/sel.2019.0018
11 Vermorel, H., & Vermorel, M. (1989). [The Sigmund Freud – Romain Rolland letters (1923–1936)]. *Revue Internationale d'Histsoire de la Psychanalyse*, *2*, 279–291.
12 Krebs, A. (2017). *Stimmung*: From mood to atmosphere. *Philosophia*, *45*(4), 1419–1436. https://doi.org/10.1007/s11406-017-9890-4
13 Nevers, P., Gebhard, U., & Billmann-Mahecha, E. (1997). Patterns of reasoning exhibited by children and adolescents in response to moral dilemmas involving plants, animals and ecosystems, *Journal of Moral Education*, *26*(2), 169–186. doi: 10.1080/0305724970260205
14 Glass, W. H. (1991). *On Being Blue*. David R. Godine.
15 Oppermann, S. (2019). Storied seas and living metaphors in the blue humanities. *Configurations*, *27*(4), 443–461. https://doi.org/10.1353/con.2019.0030
16 *The Internet Classics Archive – Meteorology by Aristotle* (n.d.). Retrieved 30 November 2022 from http://classics.mit.edu/Aristotle/meteorology.4.iv.html
17 Morton, T. (2007). *Ecology Without Nature*. Harvard University Press.

PART 2

Sea Country, Blue Country

From the postcolonial blue to the great ocean

5

SANITARY CITIZENSHIP IN THE SETTLER COLONIAL CITY

Race, health, and hygiene in interwar urban Australia

Ruth A. Morgan

In the early decades of the British colony of New South Wales, colonists danced between drought and deluge as their footprint grew from Eora Country at Warrane (Sydney Cove) to the Dyarubbin (Hawkesbury-Nepean River) of the Darug people. Off Port Jackson, the presence of the Gadigal stream that trickled into Sydney Cove had encouraged Captain Arthur Phillip to establish the first settlement there, instead of Kamay (Botany Bay), where Lieutenant James Cook had landed in 1770 (Lloyd, 1988). This waterway soon became known as the Tank Stream, after the three storage tanks that were dug into the sandstone to secure the colony's water supplies during the El Niño drought of the early 1790s. It soon became 'no uncommon circumstance', a visitor recalled, for Indigenous people to fetch water from the tanks to participate in the colonial economy (Collins, 1804).

To preserve this water source from erosion and pollution, Phillip enacted the first legislation to protect its catchment area by passing a regulation that forbade the cutting of trees within 50 feet of the stream (Collins, 1804). Phillip's successor quickly moved to prevent the Tank Stream's further pollution. Just a month after becoming Governor in September 1795, John Hunter prohibited raising pigs nearby or clearing paths to the waterway, threatening to tear down the house of the offender (Bonyhady, 2000). These orders seemingly fell on deaf ears, as he and his successors re-issued the order, and expanded its remit to prohibiting tanneries, slaughterhouses, breweries, distilleries, and so forth, with punishment including heavy fines, destruction of homes, and flogging (*HRA*, 2 May 1797). Despite these orders, the chorus of complaints about the pollution of the Tank Stream continued; by the late 1820s, it was 'doomed to become a filthy sewer' for the settlement (Houison, 1901).

On the driest inhabited continent on earth, freshwater's importance long precedes British invasion, as Indigenous peoples have actively harnessed riverine,

DOI: 10.4324/9781003365501-8

marine, and groundwater resources as well as rainfall for thousands of years, as shown by sites such as the Brewarrina stone fish traps (Baiame's Ngunnhu) on the Barwon River (Maclean et al., 2012), wooden and earth traps and weirs on the Darling (Baaka) River (Martin et al., 2022), and the World Heritage-listed Gunditjmara freshwater fishery at Budj Bim (McNiven & Bell, 2010). Water, as we have seen, also configured the earliest British incursions on the continent, as this maritime people sought out freshwater supplies, arable land, and navigable rivers to establish the farthest reaches of the Empire in what they would learn to be an extremely variable climate (Atkinson, 2016; Karskens, 2016).

Over the course of the nineteenth century, the British invaders followed a pattern of establishing their settler colonies near the confluence of fresh and saltwater courses, just as they had at Sydney Cove in 1788. Historians have charted the rise of the sanitary, networked settler city that followed, as colonial authorities responded to growing settler expectations that their water supplies would be clean and that their wastes would be removed from urban areas (see Cook et al., 2022). These expectations themselves had emerged from changing settler understandings of public health and disease, which by century's end, blended earlier environmental conceptualisations of the human body with the ascendant bacteriological order. As this chapter shows, a particular form of hydropolitics emerged in Australia's settler colonial cities, whereby access to clean water supplies became a means for social inclusion and exclusion in the settler suburbs that cut across the familiar categories of race, class, and gender, and that manifested in spatial terms.

The ways that freshwater supply has both spatially and culturally ordered *who* belongs in the Australian city, as this chapter demonstrates, reveals how settler cultures of cleanliness, health, and hygiene became embedded in the urban form of Australian cities. In these emerging settler cities, freshwater as much as land was the object of, and means of, Indigenous dispossession: its quality, availability, and accessibility vital to the settler project, whereby 'settlers sought new homes, fashioned new places of belonging out of the land they appropriated, and purposefully set about recreating its physical and ecological condition through domestic plantings' (Banivanua Mar & Edmonds, 2010, p. 4). Yet freshwater could also be a source of settler anxiety, in the form of the rivers and swamp wetlands that, in defying cadastral and sanitary order, formed 'nervous spaces' in settler cities by providing opportunity for freer movement and sustenance for Aboriginal people. These fluid terrains thus afforded Aboriginal people the means to resist their displacement, and to live in closer proximity than newcomers desired (Byrne, 2010; Edmonds, 2010; O'Gorman & Morgan, 2021).

Within the settler polity too, the condition of the streets, household, and physical body aroused considerable anxiety as to both the moral and physiological health of the colony. As the sanitary infrastructure of the settler city grew, the connection of the household to pipes, drains, and sewers facilitated further the moral association of public health and hygiene with the sanitary conduct of the occupants. In the British context, Thomas Osborne suggests,

> What was at stake was not just a Victorian fetish for cleanliness, but a strategy of indirect government; that is, of inducing cleanliness and hence good moral habits not through discipline but simply through the material presence of fast-flowing water in and through each private household.
>
> *(1996, p. 115; also Joyce, 2003)*

Household access to such fast-flowing, clean water afforded occupants access to sanitary citizenship, to borrow from Charles Briggs and Clara Mantini-Briggs (2002) in the context of a cholera epidemic in eastern Venezuela in the early 1990s. Sanitary citizenship rewards those who conform with modern medical understandings of the body and undertake hygienic practices. Additionally, sanitary citizens in Australian settler cities must be seen (by government authorities, in particular) to demonstrate their willingness and capacity to uphold and reproduce the moral sanitary order. Those deemed incapable of adopting, or unwilling to adopt, such sanitary practices were rendered 'unsanitary subjects' from whom sanitary citizens had to be protected. In settler Australia, as in Venezuela, 'The implications of being relegated to the status of unsanitary subjects were profound, affecting people's access to the political, social, and civil dimensions of citizenship' (Briggs & Mantini-Briggs, 2002, p. 11).

Examining the access to, and practices of, sanitary citizenship in settler Australian cities builds on calls to expand the purview of the blue humanities to the study of hydrosocial relations. In her 2020 study, *Hydrofictions*, Hannah Boast encourages the field to pursue waters and water cultures beyond the seas of the Anglo-American Atlantic world. Focusing on literary accounts of Israeli and Palestinian environments, Boast takes up the insights of political ecologists, who have articulated the ways that 'water is not simply acted on by humans but "is both shaped by, and shapes, social relations, structures and subjectivities"' (p. 3). Drawing on Boast's work, Matthew Henry has demonstrated through a range of literary and artistic forms the ways that 'the settler logic of elimination has long inflected hydropolitical discourse in the western United States' (2022, p. 20). This chapter argues for the closer integration of historical analyses of hydrosocial relations into the blue humanities, focusing on the expressions and consequences of sanitary citizenship in urban settler Australia.

This chapter studies the hydrosocial relations of the sanitary order in the settler cities of interwar Perth in Western Australia, Darwin in the Northern Territory, and Melbourne, Victoria. Across the continent, Aboriginal people during this period remained subject to legislation that enabled authorities to proclaim towns 'prohibited areas' for their presence and mobility, and allowed for their physical removal to spaces designated as 'reserves'. Such legal restrictions on the personal liberty of Aboriginal peoples 'reached a climax' during the period between the Great Depression and the Second World War (Rowley, 1971, p. 42). They, as well as other raced, classed, and gendered bodies, were especially subject to close sanitary surveillance through the modern state's tools of photography, enumeration, mapping, and surveys, which sought, as Matthew Newsom Kerr argues for Victorian London, 'to guide conduct of the mass by means of placing the full range of possibilities under a silent, patient, distant, and steady watchfulness' (2018, pp. 240–241). By attending to the hydropolitics of settler Australian cities, this chapter reveals the fundamental role of freshwater in shaping the social relations of the modern state.

After the First World War, concerns were reignited among the British and Australian middle classes about racial degeneration and the decline of the Empire (Murphy, 2010). Roused by these worries, reformers from an emergent class of technocrats sought to overcome social ills through the application of principles of science and reason to domestic life (Reiger, 1985). Stimulating this movement were widely held anxieties about the influence of the environmental conditions inside and outside the (middle class) home on the mental and physical health of white Australians. Meanwhile, the body and its health had come to be seen as a closed system, distinct and separate from its material surrounds. These ideas point to the paradox that historian Linda Nash has observed in modern public health: 'it insisted on the need for certain environmental changes while denying that the environment played an active role in the production of disease' (2003, p. 113). Aligned with the interests of the modernist state, advocates of these environmental reforms were mostly male members of an emerging professional middle class seeking to reshape Australian suburbs, homes, and families according to their vision of a modern settler society (Reiger, 1987).

In the suburbs of Perth, local advocates called for the improvement of parks and playgrounds to better the lives of Western Australians, particularly children (Gregory, 2009). Water was an implicit, but important tool to achieve such reform. By the end of the nineteenth century, a constant water supply had come to symbolise modern civilisation and these supplies provided the means to cultivate attractive open spaces (Taylor & Trentmann, 2011). Both indoors and out, water provided the means by which settler Australians could uphold and perform the ideals of suburban respectability, those disciplines of behaviour such as cleanliness, sobriety, and thrift. For those of more limited means, maintaining respectability meant living 'with a perpetual sense of threat', as historian Janet McCalman argues, such were its implications for 'the integrity of the personality and the family, the good regard of the neighbourhood and employers' (1998, pp. 19–20).

The expansion of Perth's reticulated water supplies had coincided with changing expectations about the appearance of the city and its suburbs. The city's town planning movement had emerged during the First World War and its influence on suburban development in Western Australia was almost immediate. The leaders of the local movement widely advocated the importance of orderly design, open spaces, and aesthetic appeal in town planning (Gregory, 2009). In 1918, the President of the Town Planning Association of Western Australia argued,

> We must do better than we have done in conserving our baby life by taking greater care of the mothers of the nation. We can augment our man power best by improving the workmen's homes, by beautifying them individually, and in the mass, by preserving the sanctity of the home and keeping the flag "one family, one house" flying by abolishing overcrowding; by getting the factories and the workers out into the environs of the city by rapid communication services to zones where land is cheaper, where air is pure, and where the children, in their garden villages, will grow up taller, stronger, deeper in the chest, freer from physical defects, happier, more likely to be stalwart effectives in the wealth-creating forces of the State, and less likely to be a burden on the community.
>
> *(Saw, 1919, p. 41)*

Attention to these details not only reflected the ideals of the international town planning movement, but was also evidence of the prevailing anxieties about the influence of the environment on the mental and physical health of suburban settler Australians.

Creating the ideal environment for the healthy moral and physical development of white Australian children was the duty of the state's citizens, particularly mothers. Citizenship, therefore, was closely related to the upkeep of the home. In March 1920, the *Daily News* declared,

> Last night practically the whole city was without water, and the sorry spectacle was witnessed, even in Hay-street West, of mothers taking their pitchers to those who possessed wells, to obtain water with which to wash their children before putting them to bed.
>
> *(p. 8)*

Reminiscent of the city's Great Water Famine of the late 1890s (Morgan, 2015), this account highlights the domestic relationship between women and water, and the ways in which gender relations shaped water use in suburban Australia in the early twentieth century. At the time of these shortages, nearly half the average Perth household's water consumption took place inside the home (Williamson, 1975).

As the *Daily News* account suggests, such episodes of water scarcity interrupted the domestic routines of Perth's housewives and domestic servants, adding to their heavy household burden (Evans & Saunders, 1992). After all, argued the local magazine *Western Homes* in 1929, they were responsible for making the ideal home: 'The parents of the West must strive to make a home of their house, and create that "Home Influence" which does not distort but beautifies; which makes *good citizens*, and not bad parasites or dangerous criminals' (cited in Gregory, 1997, p. 88). As sociologist Kerreen Reiger notes, women's daily chores had become transformed into 'scientific work of national importance' – creating good citizens (1987, p. 507). Water shortages exposed the dependence of many suburban households on reliable supplies to maintain the veneer of civility, and exacerbated suburban anxieties on the domestic frontier.

Interactions with water in early twentieth-century Australian homes and suburbs were deeply shaped by gender, class, and race. As historian Anna Haebich argues, 'Building on their racist attitudes towards Aborigines, the [Western Australian] settlers used the Aboriginal presence as a useful negative definer of what the new communities believed they stood for; for example "cleanliness" and "orderliness" versus "disease" and "dirtiness"' (1992, p. 129). Since the turn of the twentieth century, Perth's suburban expansion increasingly encroached on the lifeways of local Aboriginal peoples, whom the Western Australian government barred from the city's centre between 1927 and 1954 (Gorman, 2016). On the city's outskirts, Aboriginal peoples found places to camp and pick up work, while maintaining connections to Country. Near Galmamaanup, the swamp on the other side of the railway in Claremont, lived a group of about twenty Whadjuk Noongar. Without access to the city's growing sanitary network, they obtained freshwater by digging wells, or, as Corrie Bodney remembered, by collecting water from 'Neil's paddock close by because he had a pump there' (cited in Cook, 2016, p. 35). Another group of Aboriginal people camped each summer near a riverside beach on Freshwater Bay. They occasionally asked white residents for water, but mainly sourced 'brackish but drinkable' water by 'digging a hole in the beach sand a yard or two from the high tide mark anywhere along the beach' (cited in Cook, 2016, p. 131). Each morning before school, Robert Bropho recalled, 'There was no baths in them days for us. There was just a cat wash . . . a couple of handfuls of water thrown over your face from an old kerosene tin cut down the sides to it like a dish' (1980, p. 8).

In the northern Australian capital of Darwin, the sanitary surveillance of non-whites had already provided the government grounds to order the demolition of the city's Chinatown on the eve of the First World War (Chua, 2010). The Northern Territory's newly appointed Medical Officer of Health had inspected the conditions of Chinatown in 1912, justifying his recommendation for its demolition:

> Chinatown is [in] a dangerously insanitary state. All the conditions are present for the rapid spread of the disease viz., congestion of tenements which are dark, ill constructed and ill-ventilated; filthy living conditions, lack of proper drainage; utterly unsatisfactory condition of night-soil and refuse; a water supply which might easily become contaminated, and a general insanitary condition which lends itself, to the harbouring of vermin of all kinds and the breeding of flies and other disease carrying insects.
>
> *(Holmes, 1913, p. 4)*

Sydney's *Tung Wah Times*, a Chinese language newspaper, reacted quickly to the Officer's assessment and warned Darwin's Chinese residents to conform strictly to local health regulations (Chua, 2010). Already attuned to the precarity of Chinese communities in Sydney and Melbourne, the newspaper regularly printed translations of hygiene regulations regarding washing their produce with clean water (Boileau, 2017).

After the demolition of 15 tenements, the 1913 Administrator's report warned, 'it would be suicidal to allow [the Chinese] to continue to menace the health of the increasing white population of the same town by dwelling in overcrowded and insanitary tenements' (Gilruth, 1914, p. 8). Accompanying photographs offered a bleak assessment of the 'condemned . . . Chinese hovel', as one caption read – 'Note height of house, barely up to man's elbow. No provision for light or ventilation. No sanitary accommodation' (Figure 5.1; Gilruth, 1914, p. 56). Despite diplomatic interventions by the Chinese Consul General, only the outbreak of war slowed the demolition of Darwin's Chinatown as the government anxiously sought to populate the tropical north with Anglo-Europeans (Chua, 2010).

FIGURE 5.1 A premises in Darwin described as a 'Chinese hovel'.

Source: Public Health Board, 1914–1915, National Archives of Australia, Canberra, NT1915/1028, 7802048.

Investigations into housing conditions in Melbourne prompted the creation of new institutions to eliminate slums and provide social housing. In 1937, the first report of the Housing Investigation and Slum Abolition Board found 3,000 houses, most dating from the 1860s, to be unfit for habitation, with the city facing a shortage of between 25,000 and 30,000 low-rent houses. Middle-class commentary and photographs of these housing standards invariably extended to critical assessments of the inhabitants and their character (Figure 5.2 and Figure 5.3); as historian Tony Birch notes of slums in Melbourne's Fitzroy, they were portrayed as 'the antithesis of the decent family home environment' (2004, p. 8). Resident mothers were especially condemned as immoral and ignorant; their children would only thrive if 'rescued from the sordid surroundings and vicious influence of the slums', argued Melbourne reformer Frederick Oswald Barnett in the early 1930s (Barnett, 1931, p. 48). The Housing Commission of Victoria was established under the *Housing Act* of 1937 to address the problem, although the social stigma of the environmentally determinist reform campaigns lingered into the 1960s (Musgrove, 2003).

FIGURE 5.2 Closely scrutinised mothers, whom reformers deemed unsanitary subjects in Melbourne's interwar slums.

Source: F. O. Barnett, *c.*1935, State Library Victoria, 1777176.

FIGURE 5.3 The challenges facing a household seeking sanitary citizenship.

Source: F. O. Barnett, *c.*1935, State Library Victoria, 1777201.

The conditions of inner suburban homes in Melbourne also came under the scrutiny of Wilfred Prest, an economist at the University of Melbourne. After the 1941 Census was postponed indefinitely because of the War, the Commonwealth Department of Post-War Reconstruction, the University, and a group of local businessmen funded the Prest Survey to obtain information about housing conditions to inform future policy. Almost 8,000 households from across the metropolitan area were surveyed over a 17-month period, with interviewers recording details such as the type of dwelling, the condition of the garden and the extent of cultivated space, the location of the water closet, the presence of a sink and water tap in the kitchen, and whether the kitchen was 'ever used as a Bedroom' (Lack, 2019). The investigators encountered a broad spectrum of living conditions, such as homes that were 'well-built, attractive . . . [with a] well-kept garden', while others were 'in bad state . . . bathroom disgraceful. Kitchen and bathroom horribly untidy and dirty' (cited in Davison & Lack, 1981, p. 41). Although many of the householders interviewed for

Prest's survey indicated a desire to move, wartime restrictions and the scarcity of housing stock limited their mobility (Darian-Smith, 1990, p. 91).

The Prest Survey revealed disparities in domestic water facilities throughout Melbourne. In inner city Collingwood, nearly 60 per cent of households in the sample obtained water from an outside tap, compared to 5 per cent in middle-class Camberwell. Ninety-five per cent of houses in Collingwood and working-class Northcote had no hot water service and had to heat water for baths and clothes washing on a stove; in Camberwell the figure was nearly 60 per cent (Darian-Smith, 1990, p. 980). An investigator noted of a South Melbourne housewife that she

> had done her best to give her daughter a good start, but she wanted a house with a decent bathroom [that in the present house being outside in the wash-house]. She thought it would be a good thing if the daughter could have a bath every day – like the girls at work.
>
> *(cited in Davison & Lack, 1981, p. 42)*

Many Aboriginal people of Perth, meanwhile, were forced onto small reserves that were often situated near rubbish dumps and sanitary depots, where water supplies and sanitation facilities were inadequate (Haebich, 1992). Confined to the margins of white settlements on reserves and in camps, Aboriginal people could be better regulated and controlled under what historian Gyan Prakash has described in nineteenth-century India as 'a discriminatory sanitary order' (Prakash, 1999, pp. 132–133). Protecting the health of white suburban Australians, as in colonial India, required the containment of the putative source of disease – the bodies, habits, and homes of Aborigines. Bropho recalled that in the early 1940s at the Eden Hill camp on the city's eastern outskirts, the 'only water supply was from the local tip on the hill near the swamp' or from shallow wells they dug in the ground (1980, p. 27). In their tin billies residents collected water with 'tadpoles and the slime on the top', 'with a bit of flavour in it from human piss and human shit' (Bropho, 1980, p. 30). In more affluent Claremont, where the proximity of Aboriginal camps was a source of sanitary anxiety, some white households nevertheless employed Aboriginal people to relieve their own burden of sanitary citizenship. One resident, for instance, recalled Aboriginal women visiting his childhood home on Bay View Terrace to wash laundry on Monday, while in nearby Swanbourne, Aboriginal women also washed and ironed clothes (Bolton & Gregory, 1999; Bropho, 1980).

By the early 1950s, the local government could no longer abide the Aboriginal presence in the suburb and ordered the Galmamaanup camp to be bulldozed. Of the council's decision, the *West Australian* reported, 'the Nedlands road district was not a fit and proper place for a native camp [because] the white population was spreading in the Swanbourne area and . . . the board could not allow a "black camp in a white area"' (1951, p. 4). The camp's residents, like many other Aboriginal people that fell foul of such sanitary surveillance, were 'removed' to the Moore

River Native Settlement, over 130 kilometres north of Perth (Delmege, 2014). Writing two decades later, historian Charles Rowley could observe in his *Outcasts in White Australia* that the policing of sanitary citizenship persisted in Australian towns and cities: 'Prejudice is still commonly defended as maintaining a "hygiene" and not a "racial" barrier' (1971, p. 51).

This chapter has argued for the closer integration of historical analyses of hydrosocial relations into the blue humanities by attending to the expressions and consequences of sanitary citizenship in urban settler Australia. Focusing on interwar Perth, Darwin, and Melbourne, this chapter has revealed the ways that freshwater has been historically implicated in the social and moral fabric of settler Australian cities. In these growing cities, the availability of clean freshwater supplies was a means for authorities to govern both urban spaces as well as the raced, classed, and gendered bodies that resided in them. Practices of sanitary surveillance allowed for the exclusion and inclusion of particular bodies, according to their perceived ability and willingness to conform to the expectations of the sanitary state. As the examples in this chapter suggest, this regime provided for the rise of a discriminatory sanitary order that especially marginalised non-white bodies in order to physically and morally protect the settler colonial city.

References

'2nd May, 1797'. In Commonwealth of Australia (1914). *Historical Records of Australia, Series I: Governors' Despatches to and from England, Volume II, 1797–1800* (pp. 78–79). Library Committee of the Commonwealth Parliament.

Atkinson, A. (2016). *The Europeans in Australia: Volume 1, the beginning*. NewSouth.

Banivanua Mar, T., & Edmonds, P. (2010). Introduction: Making space in settler colonies. In T. Banivanua Mar & P. Edmonds (Eds.), *Making Settler Colonial Space: Perspectives on Race, Place and Identity* (pp. 1–24). Palgrave Macmillan.

Barnett, F. O. (1931). *The Economics of Slums*. [Masters of Commerce thesis, University of Melbourne].

Birch, T. (2004). 'These children have been born in an abyss': Slum photography in a Melbourne suburb. *Australian Historical Studies*, *35*(123), 1–15.

Boast, H. (2020). *Hydrofictions: Water, Power and Politics in Israeli and Palestinian Literature*. Edinburgh University Press.

Boileau, J. (2017). *Chinese Market Gardening in Australia and New Zealand: Gardens of Prosperity*. Palgrave Macmillan.

Bolton, G. C., & Gregory, J. (1999). *Claremont: A History*. UWA Press.

Bonyhady, T. (2000). *The Colonial Earth*. Miegunyah Press.

Briggs, C., & Mantini-Briggs, C. (2002). *Stories in the Time of Cholera: Racial Profiling During a Medical Nightmare*. University of California Press.

Bropho, R. (1980). *Fringedweller*. Alternative Publishing Co-operative.

Byrne, D. (2010). Nervous landscapes: Race and space in Australia. In T. Banivanua Mar & P. Edmonds (Eds.), *Making Settler Colonial Space: Perspectives on Race, Place and Identity* (pp. 103–128). Palgrave Macmillan.

Chua, M. (2010). The racial politics of public health in 1910s Darwin. *Journal of Northern Territory History*, *21*, 59–78.

Collins, D. (1804). *An Account of the English Colony in New South Wales: From its First Settlement in January 1788 to August 1801*. A. Strahan.
Commonwealth Housing Commission (1944). *Final Report, 25th August 1944*. Ministry of Post-War Reconstruction.
Cook, D. (2016). *'That was my home': Voices from the Noongar camps in Perth's western suburbs, Volume 1*. [Doctoral dissertation, Murdoch University]. https://researchportal.murdoch.edu.au/esploro/outputs/991005542695407891
Cook, M., Frost, M., Gaynor, A., Gregory, J., Morgan, R. A., Shanahan, M., & Spearritt, P. (2022). *Cities in a Sunburnt Country: Water and the Making of Urban Australia*. Cambridge University Press.
Darian-Smith, K. (1990). *On the Home Front: Melbourne in Wartime, 1939–1945*. Melbourne University Press.
Davison, G., & Lack, J. (1981). Planning the new social order: The Melbourne University Social Survey, 1941–3. *Journal of Sociology*, *17*(1), 36–45.
Delmege, S. (2014). From camp life to suburbia: Aboriginal housing in Perth. *Australian Historical Studies*, *45*(3), 368–387.
Edmonds, P. (2010). The intimate, urbanising frontier: Native camps and settler colonialism's violent array of spaces around early Melbourne. In T. Banivanua Mar & P. Edmonds (Eds.), *Making Settler Colonial Space: Perspectives on Race, Place and Identity* (pp. 129–154). Palgrave Macmillan.
Evans, R., & Saunders, K. (1992). No place like home: The evolution of the Australian housewife. In K. Saunders & R. Evans (Eds.), *Gender Relations in Australia* (pp. 176–190). Harcourt Brace Jovanovich.
Gilruth, J. A. (1914). *Report of the Administrator for the Year 1913*. Government Printer.
Gorman, S. (2016). Ugly town: Race, policy and people in Perth in the 1920s. *Thesis Eleven*, *135*(1), 99–114.
Gregory, J. (1997). Protecting middle-class suburbia: An ideal space for the citizens of interwar Perth. *Studies in Western Australian History*, *17*, 77–91.
Gregory, J. (2009). 'Let our watchword be "order" and our beacon "beauty"': Achieving town planning legislation in Western Australia. In R. Freestone (Ed.), *Cities, Citizens and Environmental Reform* (pp. 173–199). Sydney University Press.
Haebich, A. (1992). *For their Own Good: Aborigines and Government in the Southwest of Western Australia, 1900–1940*. UWA Press.
Henry, M. (2022). *Hydronarratives: Water, Environmental Justice, and a Just Transition*. University of Nebraska Press.
Holmes, M. (1913). Report of the Medical Officer of Health for the Year Ending 31st December 1912. *Bulletin of the Northern Territory: Health Report for the Year 1912*, no. 6. Department of External Affairs.
Houison, A. (1901). The Tank Stream. *Journal and Proceedings of the Australian Historical Society*, *1*(1), 5–7.
Housing Investigation and Slum Abolition Board (1937). *Progress (First) report, with appendices and supplements: Slum reclamation and housing for the lower-paid worker*. The Board.
Joyce, P. (2003). *The Rule of Freedom: Liberalism and the Modern City*. Verso.
Karskens, G. (2016). Floods and flood-mindedness in early colonial Australia. *Environmental History*, *21*(2), 315–342.
Lack, J. (2019). University of Melbourne Social Survey, 1941–1943: A 1981 sample. doi: 10.26193/0URT1Q, ADA Dataverse, V3. 1 ADA.QUESTIONNAIRE.50002.pdf
Lloyd, C. (1988). *Either Drought or Plenty: Water Development and Management in New South Wales*. Department of Water Resources.
Maclean, K., Bark, R. H., Moggridge, B., Jackson, S., & Pollino, C. (2012). *Ngemba Water Values and Interests: Ngemba Old Mission Billabong and Brewarrina Aboriginal Fish Traps* (Baiame's Nguunhu). CSIRO.

Martin, S., Chanson, H., Bates, B., Keenan-Jones, D., & Westaway, M. C. (2022). Indigenous fish traps and fish weirs on the Darling (Baaka) River, south-eastern Australia, and their influence on the ecology and morphology of the river and floodplains. *Archaeology in Oceania*, *58*(1), 91–114.
McCalman, J. (1998 [1984]). *Struggletown: Public and Private Life in Richmond, 1900–1965*. Hyland House.
McNiven, I. J., & Bell, D. (2010). Fishers and farmers: Historicising the Gunditjmara freshwater fishery, western Victoria. *La Trobe Journal*, *85*, 83–91.
Morgan, R. A. (2015). *Running Out? Water in Western Australia*. UWA Press.
Murphy, K. (2010). *Fears and Fantasies: Modernity, Gender and the Rural–Urban Divide*. Peter Lang.
Musgrove, N. (2003). Private homes, public scrutiny: Surveillance of 'the family' in post-war Melbourne, 1945–1965. *History Australia*, *1*(1), 8–14.
Nash, L. (2003). *Inescapable Ecologies: A History of Environment, Disease, and Knowledge*. University of California Press.
Newsom Kerr, M. (2018). *Contagion, Isolation, and Biopolitics in Victorian London*. Palgrave Macmillan.
O'Gorman, E., & Morgan, R. (2021). Fluid terrains: Approaches in environmental history. *Australian Historical Studies*, *52*(2), 141–170.
Osborne, T. (1996). Security and vitality: Drains, liberalism and power in the nineteenth century. In A. Barry, T. Osborne, & N. Rose (Eds.), *Foucault and Political Reason: Liberalism, Neo-liberalism and Rationalities of Government* (pp. 99–122). University of Chicago Press.
Prakash, G. (1999). *Another Reason: Science and the Imagination in Modern India*. Princeton University Press.
Saw, W. A. (1919). Some aspects of town planning. *Journal and Proceedings of the Royal Society of Western Australia*, *5*, 39–61.
'Swanbourne natives', *West Australian*, 4 April 1951, 4.
Reiger, K. (1985). *The Disenchantment of the Home*. Oxford University Press.
Reiger, K. (1987). All but the kitchen sink: On the significance of domestic science and the silence of social theory. *Theory and Society*, *16*(4), 497–526.
Rowley, C. D. (1971). *Outcasts in White Australia*. Australian National University Press.
Taylor, V., & Trentmann F. (2011). Liquid politics: Water and the politics of everyday life in the modern city. *Past and Present*, *211*(1), 199–241.
'A waterless city', *Daily News*, 14 March 1920, p. 8.
Williamson, D. R. (1975). Statistics of water use. *Proceedings of Seminar H_2O: Domestic extravagance*. Water Research Foundation of Australia.

6

'FROM THE VIEWPOINT OF THEIR NATIVE ELEMENT'

Diving in the colonial undersea

Killian Quigley

The trajectories of environmental humanities and social sciences research are becoming extensively submerged. Too numerous and diverse to summarise neatly here, these approaches nonetheless share a sense that underwater atmospheres engender significant – and possibly ameliorative – alternatives to 'terrestro-centric' and surficial spatialities, temporalities, and epistemologies (Anderson & Peters, 2014, p. 8). 'Blue humanities scholarship,' writes the literary historian Steve Mentz, 'removes human actors from controlling heights and plunges them into uncertainty, movement and dissolution' (2018, p. 70). Whether metaphorical or actual (or both at once), a plunge of this sort signals the rising importance of swimming and diving as objects not only *of* scholarly interest but *for* scholarly identification. The present chapter is preoccupied, in particular, by divers and diving as potent icons of knowledge-making. Writing it, I join others in seeking a richer understanding of where these icons have and have not been coming from – and of what worlds they have and have not been committed to encountering.

Concerns like these may acquire special urgency beneath colonial waters, ambiguous zones of pivotal – and poorly understood – importance for the physical and existential configuration of settler subjectivities. By plunging into and thinking from Australian seas, therefore, it may be possible to reckon with some consequential habits of submarine figuration. In 1952, *The Australian Journal* published two feature articles by the writer, critic, and political organiser Jean Devanny describing a sojourn at the Great Barrier Reef. Devanny's destination was a small island – a coral cay, to be precise – 27 kilometres off the coast of tropical north-east Queensland. To the Guru-Gulu Gungandji and Mandingalbay Yidinji Aboriginal peoples whose Sea Country encompasses the cay and its reefy surrounds, the area is known as Wunyami (Jordan, 2010, p. 149). On the terms of settler-colonial toponymy it is called Green Island, a name mooted by the navigator James Cook

DOI: 10.4324/9781003365501-9

in 1770 in the course of his first Pacific voyage (Zeppel, 2011, p. 39). Devanny relates journeying there – to what she calls 'the gem of the coral cays of the Great Barrier Reef' (1952a, p. 12) – in the company of two photographer friends, who transport an array of dive gear among their substantial equipment. During her time on the island Devanny experiments repeatedly with going below, and as she does so she discovers worlds of transcendent wonder among subsea realms that feel at once utterly novel and yet somehow like 'home' (1952b, p. 12).

More than seventy years have passed since Devanny plunged into the Coral Sea, but her marvellous immersions still read like prototypes of much of the subaqueous nature writing that has appeared in the intervening period. Here are vistas of a sort of underwater 'paradise' bedecked with lives, colours, and forms that threaten to defy the 'landlocked' imagination's powers of reckoning (Berwald, 2018, p. 5). Here, too, is the tantalising spectre of reuniting with a primordial humanness defined by its connection to, and not alienation from, oceanity. But most compelling – because most discomfiting – about Devanny's articles is a moment when their watery reveries are interrupted by persons whose presences the scenario cannot seem to assimilate. Surveying her fascinating surroundings from the seabed, Devanny receives some unexpected 'visitors', men 'brown' and 'strange' whose arrival – and whose observations *of* the author herself – are framed as disturbances of the solo diver's marinal communion (1952a, p. 14). In what follows, I approach these disruptions along three strands: the submarine erasure of Indo-Pacific labours, lives, and cultures; a solitary aesthetics of dreamlike submergence; and the strangely depoliticised construal, by *The Australian Journal* and others, of Devanny's own authorship. I ask, ultimately, what these reefy reports might tell us about who and what is taken to belong beneath the waves – and how these things might matter for all our attempts to make knowledge in, of, and with the sea.

1. Devanny's seaward trajectory

The Australian Journal, which commenced publication in 1865, was part of a relatively early crop of magazines produced in what was then the colony (and would eventually become the state) of Victoria. Notably affordable, the *Journal* strove also to distinguish itself from periodicals of foreign origin by highlighting antipodean subjects and authors. 'The AUSTRALIAN JOURNAL,' wrote then-editor Marcus Clarke in 1871, 'will publish no "original" story, the scene of which is laid elsewhere than in the Colonies, or which does not – in some way – treat of Colonial life, or subjects of Colonial interest'. The magazine was 'willing,' continued Clarke, 'to protect native industry in the matter of tale-writing, but the tales must be "Colonial", and suited for "Colonial wear", not bad imitations of the French and English imported article' (quoted in Webby, 2004). Literary scholars remember *The Australian Journal*, today, mainly for having serialised Clarke's own convict novel *For the Term of His Natural Life*, from 1870 to 1872, and for regularly printing the detective fictions and other writings of

Mary Helena Fortune, sometimes under the pseudonym 'Waif Wander' (or 'W. W.') (Webby, 2000, p. 55).

Devanny, who was born in New Zealand in 1894 and emigrated to New South Wales in her thirties, was by 1952 in the latter stages of a significant career as a novelist, polemicist, and journalist. Feminist and labour politics were foundational to her oeuvre. *The Butcher Shop* (1926), her first novel, concerned what Devanny called the 'horrible, brutal, revolting' situation of a married woman living on a remote New Zealand sheep-station. The book was temporarily banned in New Zealand, Australia, Boston, and Nazi Germany. 'Life is brutal,' reflected Devanny of this censorship, 'brutal to women, working women' (quoted in Ellen, 1999). In 1936 she released *Sugar Heaven*, a social-realist fiction of strike action among the cane-fields of north-eastern Australia, relayed from the perspectives of the workers' wives. This was, to Devanny's mind, the country's 'first really proletarian novel' (quoted in Ellis, 1999). At the time, she was an active member of the Communist Party of Australia (CPA) and a key figure, especially, in the Party's influential Workers' Art Clubs (Gardiner, 2005). These affiliations were pivotal but turbulent, and the period from the 1940s until her death in 1962 saw Devanny become expelled from, be readmitted to, resign from, and finally re-establish Party membership (Stafford & Williams, 2012, p. 1080).

Sugar Heaven was the product of an initial phase in Devanny's extended preoccupation with the colossal and substantially tropical state of Queensland, a place disparaged by not a few Melburnians and Sydneysiders as Australia's 'deep north' (Kingston, 2001). With *Cindie: A Chronicle of the Canefields* (1949), Devanny returned to the subject of the sugar industry, this time by way of historical fiction. There and in her unpublished *The Pearlers*, Devanny wrought stories informed by Queensland's economic dependence upon the labours of South Sea Islanders and Indigenous Australians – and by the intersection of those labours with the settler colony's cultures and policies of racialised exclusion (Ferrier, 2007, pp. 114–115). Meanwhile, three works of nonfiction – *By Tropic Sea and Jungle: Adventures in North Queensland* (1944), *Bird of Paradise* (1945), and *Travels in North Queensland* (1951) – elaborated a rich and affectionate, if ultimately ambivalent, portrait of the north-east Australian tropics (Ryan, 2006, pp. 74–75). In the 1950s, Devanny composed (but did not publish) *Life on a Barrier Reef Island, or, Island Interlude*, the 'autobiographical memoirs' of her sojourn at Magnetic Island (Yunbenun), a 'great sprawling hulk' lying just off Townsville, the port city that had by then become her permanent home (Devanny, unpublished, pp. 1–2).

It is an important and curious fact that scholarly histories have often narrated Devanny's growing attentiveness to tropical places in tandem with her diminishing regard for the political concerns that had animated earlier periods of her career. For her editor and biographer Carole Ferrier, Devanny's joint commitments to the workers' and women's movements had made her perversely vulnerable to marginalisation by anti-feminist members of the CPA. This experience 'diverted her,' observes Ferrier, 'into writing on natural history and about aboriginal [*sic*] customs and life' (1980, p. 28). Elsewhere, these diversions and their possible causes are simply left out of summaries of Devanny's major contributions to literature and culture. When they do feature, the late Queensland years are often characterised as a time of political retirement (Wells, 2005) if not of the appearance of a wholly 'different Devanny', this one 'full of life, curiosity and enthusiasm' and devoted to 'revelling in wild nature's infinite variety, oddity and grandeur' (Downes, 2020). A transformation of this sort has been affiliated, above all, with her engagement with North Queensland's 'reefscape', a subject entailing a 'departure from normal' – i.e., terrestrial – 'consciousness' that, if 'politically creative', was so 'in a less obvious way' (Taylor, 2004, p. 19).

A thorough accounting for the status of Devanny's politics vis-à-vis her Queensland writings lies well past this chapter's scope and expertise. These several paragraphs have sought only to familiarise readers with salient intervals in her professional history and to paraphrase a critical tendency to associate her latter works with a kind of literary apoliticism. What is particularly striking about this, for present purposes, is the coincidence – real or supposed – of a seaward trajectory with a turning-away from public life. It may bear emphasising that by pointing this out, I do not intend to confirm it as a fact of Devanny's biography. What I am remarking upon, instead, is the emergence from this milieu of a familiar trope of certain marinal imaginaries: to go to – and especially under – sea is sometimes to attain a sort of exemption from the perturbations of worldly affairs. How this motif has been working is a matter of unusual pertinence not only for describing a literature of the submarine but for reckoning what a sea-going humanities may, and ought to, do – as well as what it may sometimes fail at doing.

2. The submarine world

A signal instance of Devanny's depoliticisation takes shape upon the pages of *The Australian Journal*. A blurb-cum-biographical note to the first of the Green Island articles foregrounds her being 'one of the adventurous minority' – and one of 'very few women' – who have descended in a diving bell. It explains subsequently that despite being from New Zealand, Devanny has but for one 'trip abroad' been resident in Australia for decades. This unnamed voyage was presumably the one she undertook in 1931, at the behest of the CPA, to Germany and the Soviet Union; in Berlin, she was Australian delegate to the conference on Workers' International Relief (Ferrier, 2001, pp. 117–118). On *The Australian Journal*'s view, her career has encompassed 'more than twenty books, including several travel works and a number of novels'. These go unenumerated save for *Bird of Paradise*, *Cindie*, and an ongoing 'book on Magnetic Island' (Devanny, 1952a, p. 13). An author's portrait accompanying this text presents Devanny as poised, scholarly, and unsubmerged. As it happens, the same headshot had featured several years earlier alongside a stirring op-ed in the CPA's *Tribune* newspaper. There, Devanny advocated on behalf of the rights of Torres Strait Islanders to 'the ownership and control of their territory and territorial waters and all that these contain' and railed against the 'farcical' policies of the Queensland government in this regard (1948, p. 4). Through its connections and emplacements, the photo quietly amplifies the strangeness of *The Australian Journal*'s selective biography.

The diving bell that enables Devanny to 'watch the citizens of the submarine world' swim by (1952a, p. 13) belonged to her friend Bruce Cummings, whose underwater camerawork at Green Island had featured in the filmmaker Noel Monkman's 1938 adventure *Typhoon Treasure* (Raymond, 1986, pp. 20–21). At the island, this 'submersible studio' is only one technology of visualisation among others: during her first week there, Devanny spends hours 'in Bruce's dinghy, gazing down through his glass float at the occupants of the underwater meadows of dugong grass and coral pools'. These watchings and gazings partake in turn of a broader ambience of optical charisma. Of a mercurial summer evening, to look back towards the mainland from Green Island is for Devanny to receive a sequence of extraordinary impressions: 'an inimitable pageantry of colour', to start, followed by 'a few minutes of atmospheric clarity in which distant objects loomed fantastically twice their size' (1952a, pp. 12–13). The general sense here, is of a cay-ward trip unveiling a host of exquisite and unprecedented sensations, on and above watery surfaces as well as below them, as if effecting a complete refreshment of the traveller's aesthetic faculties.

Bruce's bespoke diving bell – or 'diving cylinder' – is on Devanny's account an ingenious (if astonishingly weighty) contraption, from the configuration of its air-pipes to the 'four large windows of special optical glass' that facilitate its views. Conditions being unconducive to a deep-water plunge, arrangements are made for a nearshore descent off Green Island's main jetty. Going down there might mitigate the shock of Devanny's 'new experience of submarine observation', and in any case the 'several fathoms of water' obtaining around the jetty at high tide (between 7 and 8 metres, perhaps) would make for no insignificant depth. Frances Cummings, Bruce's wife and a noted subsea photographer in her own right, dons the 'earphones' that will permit her to receive communications from below. Meanwhile, the novice diver enters the bell and the lid is 'screwed down' above her. Devanny's confinement is productive, first, of 'an attack of nervousness' and feelings of acute isolation and loneliness – a distinctive species of claustrophobia, deriving perhaps from the situation of being contained in an apparatus of submersion while remaining temporarily unreceived by the waves. But this crisis only heightens the transporting pleasures to follow: 'as I was lowered away,' she writes, 'the amazing nature of the new experience effaced all consciousness of self' (1952a, p. 13).

What follows is a catalogue of splendid novelties, caringly and exuberantly described. As she drops, Devanny registers the jetty's piles 'thickly encrusted with fully expanded corals, asway like masses of flowers' and adorned in turn by 'tiny silvery-blue iridescent fish, round as a florin'. When the bell settles on the 'rainbow-hued sand' of the lagoon floor, its occupant's gaze takes in an environment that cannot but exalt what it contains: 'even the sardines,' Devanny reflects, 'took on glamour from the viewpoint of their native element'. Having won this writer's attention, the sardines acquire unusually vivid shapes through language that conveys not only their colours but their complex behaviours, severally and individually construed. What might have seemed from shore 'a dark uninteresting wad' becomes, at depth, an incitement to extended delineation, a sort of unanticipated hallowing of small but brilliant submarine lives. Of the corals that clad the jetty Devanny counts seven different species, their hues 'pink, salmon, green, ivory and heliotrope' and their anatomy – polyps, digestive chambers, tentacles fringing mouths 'like petals' – intelligible through optical glass. Onward the exhibition shifts, through the quiet drama of sardines scattering at the arrival of 'prowling trevalli' and on to the 'remarkable manœuvres' of razor fish. Diving-belled, Devanny has become involved in 'wrapt [*sic*] contemplation' of a 'dream world' lying 'upon the sunlit sands' just 'twenty-five feet below' (1952a, p. 14).

The oneiric is a perennial trope of subaqueous literature and art in general, and of reefy imaginings in particular (Cohen & Quigley, 2019, pp. 4–5; McCalman, 2019, pp. 131–133; O'Hanlan, 2019, pp. 139–145). Decades after Devanny's dive, the journalist Elizabeth Kolbert journeyed to One Tree Island, a coral cay located in the far southern reaches of the Great Barrier Reef. Of her first experience snorkelling there, Kolbert recalled witnessing a 'swirl of life' so spectacular as to seem 'unreal'. Attempting to 'keep a mental list' of the constituents of that swirl wound up feeling, she wrote, 'like trying to catalog a dream' (Kolbert, 2014, p. 139). Devanny's passage from consciousness to raptness is by contrast exceptionally productive from the naturalist's point of view: she manages not only to identify the lives she sees but to characterise their movements, measure their proportions, and even render analogies for their appearances (the razor fish is 'shaped like a Bengal razor' (1952a, p. 14)). That she employs a bell where Kolbert is immediately immersed is of course a pivotal difference between their forms of contact. What impinges, nevertheless, on Devanny's observatory – and on the loosened self that inhabits it – when her immersed reverie is suddenly interrupted is the subject of the proceeding section.

3. These strange men

'There's a boat coming down.' With these words, conveyed into the diving bell by means of its two-way telephone, Frances's voice breaks Devanny's sea-bottom spell. 'A queer looking boat,' her voice continues. 'Can you hear it?' (Devanny, 1952a, p. 14). Submersed in an element where sound travels as much as five times faster than it does topside, a human diver is susceptible to the discombobulating potentiality of sensing a noise strongly but becoming confused as to its source (Merchant, 2011, pp. 226–227). At Green Island, Devanny quickly hears the 'rumblings' of propeller and engines and feels spurred, by them and by the increasing agitation of the water around her, to ask to be pulled up. She is lifted, but only slightly, and the bell is 'left hanging' only a short way off the bed. 'You're going to have some visitors,' comes Frances's next transmission. 'Some people up here are intrigued by the cylinder' (Devanny, 1952a, p. 14).

No sooner does this news reach Devanny but 'an immense frog-like shadow' enters her field of vision. What the shadow heralds is a new scene for the bell's optical glass, albeit one that its spectator will find less edifying.

> Then the sea became alive with men. Strange men, brown men, of a type I had never seen before. One kicked himself up to a window and stared at me through red-rimmed goggles. Another, a large fat man, landed on the sand, dug his heels in with a tremendous shove, worked his arms like a flail and thrust his face against the glass. Like a cork he bounced up again. Others followed. Their cheeks ballooned as they stared, tapped the windows and shot up. I became anxious to go up. Who were these strange men?
>
> *(Devanny, 1952a, p. 14)*

Unlike the article's intervals of naturalist enthusiasm, this is a reckoning with liveliness and strangeness that inclines towards not expatiative fascination but terse anxiety. By addressing themselves to the glass, the goggled man, his plump companion, and the others half-echo the few bold sardines who had earlier 'peeled off' from their shoal's 'colossal column of drifting silver' to approach the bell and return its resident's look. But instead of provoking novel forms of admiration, the scopic nexus of seeing and being seen underwater lends these 'brown men' a quality, for Devanny, of enhanced oddness. The confusion this entails is more than interpersonally awkward; it bewilders the subsea atmosphere so far as to compel the writer to complete her exit therefrom (Devanny, 1952a, p. 14).

Having been hauled up, Devanny is informed that the men arrived aboard 'a New Guinea river boat' making transit to Cairns for engine-work. Its skipper, a 'white Australian', explains that 'his Papuan crew' have been growing 'restless' for lack of fish; amidst their group, Devanny remarks in particular upon 'a young woman, with arms, legs and chin one smother of tattoo marks'. As will be obvious to readers from or familiar with the Coral Sea, to construe these 'Papuans' as making an incongruous – 'queer looking' – display is to burlesque the region's real historical identities and mobilities (Devanny, 1952a, p. 14). Those prominently include familial, cultural, and political relations among (for instance) Pasifika labourers blackbirded for work in the Queensland sugar cane industry; Torres Strait Islanders whose lands and waters eventually became home to many of those labourers; coastal Papuans who fished, traded, and formed community in the Torres Strait and the Coral Sea, among other places; and Aboriginal people whose Country – including their Sea Country – lay and still lies across the north-eastern reaches of the Australian continent (Lowah, 1988). Moreover, the idea that the Cairns-bound crew make strange fellows below the surface of Indo-Pacific waters rings remarkably bizarre once considered in view of the many economically pivotal and indeed famous subsea exploits of brown and black persons involved in the region's marine industries (Quigley, forthcoming). What is odd, that is to say, is not that a Coral Sea dive would furnish an encounter of the kind Devanny experiences, but that she, her reader, or we would represent such an encounter as 'strange'.

FIGURE 6.1 *The Australian Journal*, February 1, 1952.

It is worth acknowledging that by examining this incident at some length I am writing aslant the structural protocols of Devanny's articles, which actually pay it fairly short shrift. After a very few words describing the Papuan sailors' stay, the reader is ushered hastily on – first to an account of a 'special show' the fish at Green Island give after sundown, and then to a subsequent dive, where Bruce makes a film about anemones (Devanny, 1952a, pp. 14–15). But if the episode is narratively marginalised, *The Australian Journal* has made the surprising and significant decision not only to illustrate it but to display it on the issue's front cover (Figure 6.1). That depiction reflects an alarming choice to replace Devanny's brown divers with three white men. These latter are so similar as to look, uncannily, like clones of one another, but bearing the articles' historical moment in mind, it is tempting to interpret them as evoking Jacques Yves Cousteau, Frédéric Dumas, and Philippe Tailliez, the so-called Mousquemers whose submarine heroics had lately made them internationally famous (Bright & Kimmey, 2021, pp. 5–6). Then again, if Cousteau and his collaborators' undersea aesthetics was really, substantially committed to expressing the perceptual idiosyncrasies of oceanic realms and lives

(Cohen, 2022, pp. 107–120), *The Australian Journal* neutralises such distinctiveness in pursuit of an ironically obfuscatory type of clarity.

Whomever their models might be, the cover's white men erase the persons Devanny encountered while imbuing the meeting with a specific erotic logic. The ambivalence that actually defined the Green Island scene is substituted for by a contingent convention of underwater looking, one that seems to practically prefigure the scene arranged and lampooned by the feminist artist Hannah Wilke in her film *Hello Boys* (1975), where female bodies and aquatic realms are made simultaneous objects of masculine gazing (Firebrace, 2016, p. 36). Devanny's goggle-eyed avatar, meanwhile, is portrayed as blonde and seemingly young, and if these alterations are less glaringly problematic than the others aforementioned they are not trivial for being so. A dive at the Great Barrier Reef had already occasioned, for *The Australian Journal*, the selective presentation of its author's career, a career that we need not construe as faultless in order to recognise its achievement and frequent courage. Devanny's exoticising regard for the brown woman and men she meets, at the surface and on the seabed, conduces to a distorted sense of the tropical undersea. At the same time, her error is subsumed within a colonial magazine's vision of reefy splendour, a vision that is at pains to make space for diving cylinders, white men and women, and a token handful of other-than-human oceanic lives – but has little room, in the end, even for the author of the spectacle.

* * *

On a Saturday morning in December 2022, I made my way to Reef Fleet Terminal, Cairns and boarded a large tourist boat bound for Green Island. Of the two hundred or so passengers that made the transit that morning, as well as of however many guests were staying over (for between $400 and $500 nightly) at the resort at the time, I was, to my surprise, the sole diver. Immediately after arriving, I walked to the SCUBA shop to gear up. 'Do you need a cameraman?' a staff-member asked me politely. I said thanks but no, wondering the while what images of the place the people working there find themselves asked to produce by tourists like myself. Then I pulled on my stinger suit and dive kit, and we motored off – a skipper, a guide, and me – to a site they called 'New York', located some few hundred metres off the island's west end. Its warm, turbid waters divulged some extraordinary scleractinian corals, dazzling sponges, many wonderful fish, and one of the largest giant clams I'd ever seen. Dozens of hexagonal metal frames stood from the sea-bottom, parts of a 'reef regeneration' project that aims to furnish a 'new stable location' for corals subject to 'accumulating environmental stressors' (Green Island reef rehabilitation project). Between plunges, I lay in the shallows a little way from a green turtle as it munched on seagrass, apparently unbothered by my presence, or by anyone else's.

Just a little more than seven decades since Devanny's dive, the extinguishment of the Great Barrier Reef's multispecies brilliance is for many a starkly worrying – if not certain – prospect (Quigley, 2021). At the same time, a colonial aesthetic of reefy wonder has become the object of extended scrutiny (Elias, 2019). These concerns intersect, in Devanny's descent, with the question of what human presences the seascapes of naturalist and recreational diving have been more and less capable of assimilating. This brief chapter has presented the scene of submergence as one worth reckoning for the complexities it really expresses, especially when those complexities have been obscured through influential tropologies of the who, what, and where of going-beneath. If submarine trajectories have tended, in certain significant cases, towards the occlusion of human sociability and politics, a critical ocean studies asks what alternative practices of plunging might carry our descents beyond solo rapture (DeLoughrey, 2017). What have always lain past the horizons of the settler-colonial Australian undersea are – in the terms of the Aboriginal writer and organiser Cherissma Blackman – some far more distinguished ethical, aesthetic, and socio-ecological stories (To, 2022). The textures and contours thereof well exceed my descriptive capacities, but the closing words of this essay acknowledge their enduring and resurging presence – and hopefully anticipate their becoming better-known among waters integrally learned, and submersions carefully undertaken.

References

Anderson, J., & Peters, K. (2014). Introduction: 'A perfect and absolute blank': Human geographies of water worlds. In J. Anderson & K. Peters (Eds.), *Water Worlds: Human Geographies of the Ocean* (pp. 3–19). Ashgate.

Berwald, J. (2018 [2017]). *Spineless: The Science of Jellyfish and the Art of Growing a Backbone.* Black Inc.

Bright, D. A., & Kimmey, R. (2021). Does anything dive? Diving beyond the metaphor. *Techniques & Culture* [Online]. Suppléments au n°75. https://doi.org/10.4000/tc.15034

Cohen, M. (2022). *The Underwater Eye: How the Movie Camera Opened the Depths and Unleashed New Realms of Fantasy.* Princeton University Press.

Cohen, M., & Quigley, K. (2019). Introduction: Submarine aesthetics. In M. Cohen & K. Quigley (Eds.), *The Aesthetics of the Undersea* (pp. 1–13). Routledge.

DeLoughrey, E. (2017). Submarine futures of the Anthropocene. *Comparative Literature, 69*(1), 32–44.

Devanny, J. (1948, 10 March). Torres Strait Islanders need a new deal. *Tribune, 4.*

Devanny, J. (1952a, 1 February). Down in a diving bell. *The Australian Journal,* 12–15.

Devanny, J. (1952b, 1 March). Denizens of the coral seas. *The Australian Journal*, 12–15, 30.

Devanny, J. (Unpublished). *Life on a Barrier Reef Island, or, Island Interlude* [Manuscript]. James Cook University Library. https://nqheritage.jcu.edu.au/756/

Downes, L. (2020). Summary. Life on a barrier reef island or island interlude. James Cook University. https://nqheritage.jcu.edu.au/756/

Elias, A. (2019). *Coral Empire: Underwater Oceans, Colonial Tropics, Visual Modernity.* Duke University Press.

Ellen, M. R. (1999). *The Butcher Shop* (1926). In L. Sage, G. Greer, & E. Showalter (Eds.), *The Cambridge Guide to Women's Writing in English.* Cambridge University Press. Credo Reference.

Ellis, C. (1999). *Sugar Heaven* (1936). In L. Sage, G. Greer, & E. Showalter (Eds.), *The Cambridge Guide to Women's Writing in English.* Cambridge University Press. Credo Reference.

Ferrier, C. (1980). Jean Devanny's Queensland novels. *LiNQ, 8*(3), 20–30.
Ferrier, C. (2001). Jean Devanny in Australasia and the Soviet Union in the 1920s and 30s. *New Zealand Slavonic Journal, 35*, 113–126.
Ferrier, C. (2007). Jean Devanny, 1894–1962. *Kōtare, 7*(1), 99–118.
Firebrace, W. (2016). *Memo for Nemo*. AA Publications.
Gardiner, A. (2005). Communist Party of Australia (literary activities). In E. Benson & L. W. Conolly (Eds.), *Encyclopedia of Post-Colonial Literatures in English* (2nd edn). Routledge. Credo Reference.
Green Island reef rehabilitation project. Great Barrier Reef Marine Park Authority. https://www2.gbrmpa.gov.au/our-work/field-management/green-island-reef-rehabilitation-project
Jordan, D. (2010). Environment and colonial shadows: Green Island 1932. *LiNQ, 37*, 142–158.
Kingston, B. (2001). Queensland. In G. Davison, J. Hirst, & S. Macintyre (Eds.), The Oxford *Companion to Australian History*. Oxford University Press. Oxford Reference.
Kolbert, E. (2014). *The Sixth Extinction: An Unnatural History*. Picador.
Lowah, T. (1988). *Eded Mer (My Life)*. Rams Skull Press.
McCalman, I. (2019). Encountering living corals: A nineteenth-century scientist and artist reveals the underwater realm. In M. Cohen & K. Quigley (Eds.), *The Aesthetics of the Undersea* (pp. 111–123). Routledge.
Mentz, S. (2018). Blue humanities. In R. Braidotti & M. Hlavajova (Eds.), *Posthuman Glossary* (pp. 69–72). Bloomsbury Academic.
Merchant, S. (2011). Negotiating underwater space: The sensorium, the body and the practice of scuba diving. *Tourist Studies, 11*(3), 215–234.
O'Hanlan, S. T. (2019). The shipwreck of reason: The surrealist diver and modern maritime salvage. In M. Cohen & K. Quigley (Eds.), *The Aesthetics of the Undersea* (pp. 137–154). Routledge.
Quigley, K. (2021). Caring for colour: Multispecies aesthetics at the Great Barrier Reef. *Queensland Review*, *28*(2), 82–93.
Quigley, K. (forthcoming). A grotesque and inhuman world: pearl shell, diving knowledge, and the apparatus of race. In C. S. Perez (Ed.), *For the People, Land, and Sea: Environmental Relations in Oceania and the Pacific Islands*. University of Washington Press.
Raymond, R. (1986). *Starfish Wars: Coral Death and the Crown-of-Thorns*. South Melbourne: Macmillan.
Ryan, S. (2006). The holiday-maker's happy hunting ground: Travel writing in Queensland, 1860–1950. *Queensland Review*, *13*(1), 63–77.
Stafford, J., & Williams, M. (Eds.) (2012). *The Auckland University Press Anthology of New Zealand Literature*. Auckland University Press.
Taylor, C. (2004). Threshold to fulfillment: The Barrier Reef writings of E. J. Banfield and Jean Devanny. *Antipodes*, *18*(1), 18–23.
To, C. (2022). Songs of the Reef – Crouch End Festival Commission. *Resonate Magazine*. Australian Music Centre. https://www.australianmusiccentre.com.au/article/songs-of-the-reef-crouch-end-festival-commission
Webby, E. (2000). Colonial writers and readers. In E. Webby (Ed.), *The Cambridge Companion to Australian Literature* (pp. 50–73). Cambridge University Press.
Webby, E. (2004). *The Australian Journal: History*. AustLit. https://www.austlit.edu.au/austlit/page/C498899
Wells, J. (2005). Devanny, Jean (1894–1962). In E. Benson & L. W. Conolly (Eds.), *Encyclopedia of Post-Colonial Literatures in English* (2nd edn). Routledge. Credo Reference.
Zeppel, H. (2011). Managing tourism on Green Island, Great Barrier Reef Marine Park: Conservation, commerce, and compromises. In J. Carlsen & R. Butler (Eds.), *Island Tourism: Sustainable Perspectives* (pp. 38–53). CABI.

7

THE 'BLUE TURN' IN CONTEMPORARY ART

Assembling blue methods of research-creation

Jacqueline Chlanda, Léuli Eshrāghi, and Peta Rake

Taking place on the lands and waters of local Yuggera and Turrbal First Nations, in 2022 we staged two exhibitions at the University of Queensland Art Museum that invited audiences to consider how we might be able to think together with ocean species and places to speculate on collective futures. *Oceanic Thinking (Season One)* opened in the southern hemisphere's late summer, followed that winter by *Oceanic Thinking (Season Two)*. Through the lens of contemporary art, these exhibitions sought to provide a foundation for audiences uninitiated in critical ocean studies, critical Indigenous studies, or the blue humanities. These ways of thinking about oceans through socio-political and ethical frameworks include both humans and what Donna Haraway (with others) calls the 'more-than-human' – the critters, ecologies, geographies, and more that are entangled with us, and without us (Haraway, 2015; see also Iovino & Oppermann, 2014).

The *Oceanic Thinking* exhibitions were the first in a focused multi-year research and programming arc titled *Blue Assembly*. Exhibitions, public dialogues, research-creation residencies, a year-long online journal, and an interdisciplinary summer workshop all make up the ongoing work of *Blue Assembly*. The project is dedicated to plural epistemologies, relationalities, and ontologies as a means of charting ways through contemporary climatic challenges, but also seeks to think beyond the contemporary moment.

Importantly, neither iteration of *Oceanic Thinking* was framed through the Anthropocene – the current epoch of irreversible, destructive human impact on the planet – due to the term centring on the ancient Greek 'anthropos' for man but in practice referring solely to the constructions of modernity as exclusively arising out of Europe. Rather than home in on the context of the Industrial Revolution or on the Eurocentric whiteness of capitalism and ecological exhaustion, we sought to render plural and complex understandings of this moment (Haraway, 2016).

DOI: 10.4324/9781003365501-10

This being said, many of the works did signal a concern with impacts of the Anthropocene or 'Capitalocene' – Jason Moore's (2017) term, which re-periodises this epoch of environmental crisis as rooted in the earlier advent of capitalism rather than the Industrial Revolution – alongside the politics and histories of various colonisations. But the oceans present across these works were bigger than this so-called epochal moment and its structural frameworks; across the exhibition, artists charted an entangled pelagic diaspora, speaking to contemporary ways of being and living. Some of the artworks asserted the continuity of customary practices, but all were acts of resistance to the dominant narratives of so-called progress and accumulated capital. In this chapter, we draw out the broad frameworks that inform *Blue Assembly*, before examining a number of the artworks from the two exhibition seasons of *Oceanic Thinking* made by Australian-based artists from a range of diasporic backgrounds. These works enact, as case studies, what Melody Jue refers to as the 'cognitive estrangements' – disruptions to normative, terrestrial modes of understanding – that oceanic thinking allows (2020, p. 84).

Throughout this chapter, we use the term 'Great Ocean' in place of 'Pacific Ocean'. This is inclusive of Australia and further rim territories and is an English translation of relational and expansive terms in reo Māori, gagana Sāmoa, 'ōlelo Hawai'i, Hakö, and A Tinata Tuna, among other Indigenous languages of the region. We use the term 'research-creation' to describe artists' practices which encompass material, aesthetic, and intellectual inquiry through research, critique, debate, and appraisal.[1]

Making visuality through tidalectic curating

For both seasons of *Oceanic Thinking* curatorial decisions were motivated by a desire to tell multiple, complicating stories about oceans and other blue spaces. The exhibitions were underpinned by what Stephanie Hessler refers to as 'tidalectic curating' (2020, p. 251): a methodological approach to exhibition making that takes oceans as both topic and framework, so that they are both addressed – conversed with and about – and thought through. Employing poet Kamau Braithwaite's generative concept of 'tidalectics', Hessler analyses contemporary curatorial practices guided by 'natural processes such as the ebb and flow of the tide', which allow, among other things, thinking beyond national borders (2020, p. 267). Curators working in this vein take the ocean as 'a model to accommodate both change and unpredictability, to sway back and forth between, and ultimately to transcend numerous disciplines and to invoke performativity, both in the exhibition making process and in the audience experience'. Aptly, she asserts that '[t]idalectic curating does not speak about, but through, with and "nearby"' (Hessler, 2020, p. 258).

Thinking 'with' instead of 'for' nature and non-human entities has gained traction as a curatorial and artistic methodology in recent years, and is a key tenet marking a blue, or aquatic, turn in contemporary art and research-creation more broadly (Morse, 2022). It is important to recognise the leading work of artists in proposing transoceanic and situated approaches to plural cultural memories, futurities, and solidarities. We name here a few of the many we have been informed and inspired by: *Reclamation: Taloi Havini* (2020), co-commissioned and presented by Artspace and Sydney Festival; *Mel O'Callaghan: Centre of the Centre* (2020), co-commissioned by Artspace, University of Queensland Art Museum, and Le Confort Moderne; *Santiago Mostyn: Dream One* (2022), commissioned by Södertälje Konsthall; *Shannon Te Ao: Ka Mua, Ka Muri* (2020–1), commissioned by Remai Modern; and the 2021 exhibition *Tree Story* at Monash University Museum of Art (MUMA) in Melbourne, curated by Charlotte Day and Brian Martin, which was conceived as a 'forest of ideas' (Day & Ratcliffe, 2021). *Tree Story* brought together artists' projects from around the world to connect critical environmental issues, such as recurring, worsening forest fires and salinating river systems, to a grounding in Indigenous ways of knowing, particularly First Nations' living practices of seeing eucalypts as tree kin. Practices of caring for Country are consonant with this thinking 'with' trees; such networked relations emphasise the importance of their protection, as a human as well as a more-than-human concern and responsibility. *Oceanic Thinking* employed a similar approach to *Tree Story*, thinking with oceans through the invested practices of climate activism and Indigenous custodianship of Country. This included a visit to sea grass forests in Quandamooka Country to the east of Brisbane in late 2021 as part of the exhibition development. As with *Tree Story*, the didactic texts on gallery walls were framed largely around transdisciplinary concerns of climate pasts and futures, rather than through distanced art historical frameworks. These texts foregrounded pelagic and diasporic communities' relationships to blue spaces, as a way to unsettle assumptions about positionalities and our duties to more-than-human kin.

The move towards thinking 'with' disrupts curatorial approaches to the Anthropocene that, as Hessler puts it, 'fail to consider the exhibition as affecting the field that it describes . . .' (2020, p. 258). Exhibitions that hold themselves at a distance from their topics – which position themselves as *about*, but not informed *by* – are inadequate in the face of intersecting emergencies.[2] Hessler urges a much more invested model of exhibition making. Central to the conceptual framework of *Oceanic Thinking* was the complex and rewarding process of ensuring the entire exhibition-making process through to opening and visitation was carbon neutral. We are the first museum or gallery in the state of Queensland to have an exhibition certified as carbon neutral, and only the fourth in Australia.[3]

It was similarly critical to *Oceanic Thinking* that the environment and the spectre of climate change were thought-through cultural practices that refute the dominant (Western) binary of nature–culture. Fijian–Italian researcher Talei Mangioni reminds us that climate change discourse remains largely 'in the distinct arena of big science rather than that of Indigenous thinking, the humanities or arts' (2019). For Mangioni, because climate change is a continuation of the violence and dispossession of colonisation, climate justice is inseparable from decolonisation, recognising that '[o]ceanic cosmology is deeply intertwined with the environment' (2019). Artist Taloi Havini reflects the way in which this closeness between Great Ocean peoples and their environments shapes their perspectives: '[the] West looks down on the world and on the ocean. They look down on the land. Whereas we look out and from within' (Havini & Taumoepeau, 2020). Australian Indigenous researcher and Goenpul woman of the Quandamooka people (Moreton Bay) Aileen Moreton-Robinson expresses a similar view when she describes the 'ontological relationship to land' (2015, p. 11) that is constitutive of Australian First Nations peoples' subjectivities. Hessler's useful concepts of 'through, with and "nearby"' (2020, p. 258) are preceded by the world views that Mangioni, Havini, and Moreton-Robinson reflect, and the works discussed throughout, while not all by Indigenous artists, rely on a cultural, historical, and ontological imbrication with the natural world. It is important to the ongoing work of *Blue Assembly* to firmly acknowledge this thinking as part of, not above, the Majority World.[4]

In what follows, we look at the work of four artists included in the *Oceanic Thinking* exhibitions whose work exemplifies the 'blue turn': collaborative sibling duo Sancintya Mohini Simpson and Isha Das, Izabela Pluta, and Angela Tiatia. All are Australian based but from diasporic backgrounds. Their works focus on oceanic waters as peopled, as relational sites of family history and systemic exploitation, as a means to rethink vision and thus critically engage with the way oceans are made visible to us, and as a cosmology that transcends colonial occupation and exploitation.

Navigating Dark Waters

Brisbane-based siblings and artistic collaborators Sancintya Mohini Simpson and Isha Das are the descendants of indentured labourers sent across the ocean from South Asia under the British Raj to work on sugarcane plantations in the colony of Natal (now the South African state of KwaZulu–Natal). The artists' familial histories of indenture echo the experiences of South Sea Islander communities, whose ancestors were trafficked to sugarcane and cotton plantations along the length of the northeast Australian state of Queensland in a practice referred to as blackbirding, something Simpson is explicit about in her public discussions of her practice.[5] Her work in *Oceanic Thinking (Season One)*, including her collaborative installation made with Das, offers ways through which to understand the blue through an intergenerational lens of caring for ancestors and descendants, tracing the gaps in cultural transmission along the routes taken by the countless ships rendering the oceanic passes of this planet as कालापानी, kālāpānī, Dark Waters (Simpson & Das, email correspondence, 10 February 2022).

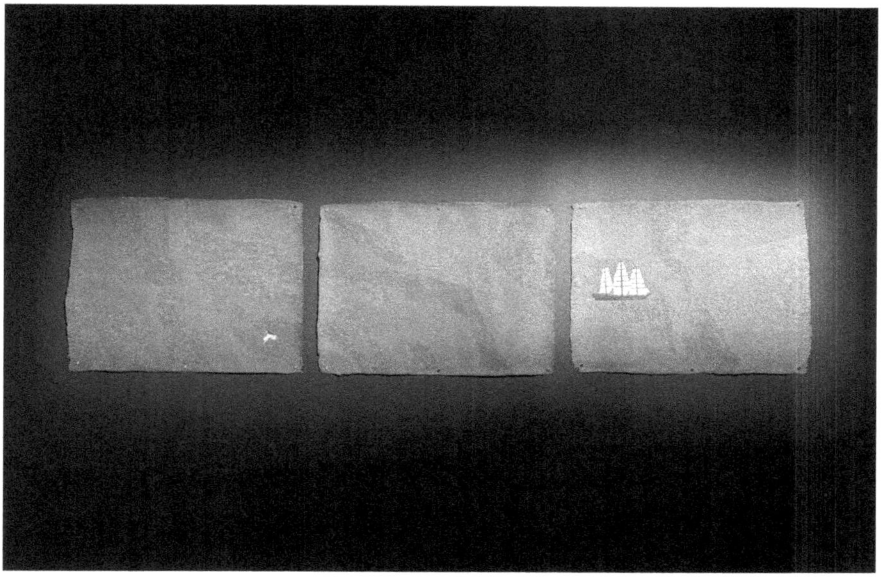

FIGURE 7.1 Sancintya Mohini Simpson, *Tāpū*, 2022. Installation view, *Oceanic Thinking, 2022*, UQ Art Museum.

Source: Courtesy of the artist and Milani Gallery, Brisbane. The artist would like to thank Isabel Wengert. Photo: Marc Pricop.

One work, *Tāpū* 2022 (Figure 7.1), is a delicate, exacting watercolour and gouache triptych on handmade wasli paper. In this work, Simpson continues her project of upending the historical privilege and conventions of Indian miniature painting in northern India and Pakistan which still reflects dominant gender, caste, and class positions in its representations. She instead focuses attention on the experiences and lives of two protagonists: a casteless woman adrift in an oceanscape and the figure of a colonial tall ship. The woman is submerged under the waves of a tumultuous sea, seemingly cast overboard, and the tall ship continues on out of the frame (Figure 7.2). The artist has shared with us that 'Tāpū' has meaning across many languages of South Asia – Sanskrit, Marathi, Hindi, among others – understood variously as a taboo, an island, a body surrounded by water on all sides, or the exiling of a person from the mainland. Here, in relation to indentured labour, Simpson takes a little known South African Bhojpuri definition: 'anyone who disappeared from the village (absconded, lost, or murdered) was said to have gone to Tāpū, so there was an evil association' (Tinker, 1988, p. 198). The slippage of language here is important, as Tāpū took on new meaning through the forced displacement of communities, across oceans, to the sugarcane fields of Natal (Tinker, 1988, p. 198).

FIGURE 7.2 Sancintya Mohini Simpson. *Tāpū* (detail), 2022; watercolour and gouache triptych on handmade wasli paper.

Source: Courtesy of the artist and Milani Gallery, Brisbane. The artist would like to thank Isabel Wengert. Photo: Marc Pricop.

In *Poetics of Relation*, Édouard Glissant (1990/1997) positions the abyss – the slave boat, the ocean, lost homelands behind, and unknown lands ahead – as a kind of beginning, the shared relation to which engendered a new people and knowledge. Glissant's pelagic perspective as a descendant of the transatlantic trade in the still-colonised francophone isles of the Caribbean is important in recognising the emotional labour implicit in, and cultural specificity of, the works Simpson and Das have produced together. They honour the complexities of forced migration, matrilineal memory, and trauma, always cognisant of thinking through their contemporary responsibility and relation to these histories. Glissant sets out a schema for identity based on relations, with a particular attentiveness to the kinds of identity generated through different forms of nomadism. He distinguishes between Western, colonial, invading ('arrowlike') nomadism and the idea of errantry (Glissant, 1997, pp. 11–22). The former arrests movement in the establishment of nations, setting down roots which both consolidate and expropriate the form of identity that the nation encodes. According to Glissant, this is a totalising, dualistic means of engendering identity. Against this, he offers errantry. For the errant, roots are rhizomatic (a concept he borrows from Deleuze and Guattari [2004]) – they still signify rootedness, but this is an open, mobile, multiple form of relational connectedness as identity (Glissant, 1990/1997). Simpson's poem 'Gone to Tāpū' signals this kind of thinking: 'Memories in the body/ Of water that carried my people/ And your people' (2022, p. 54, lines 1–3). A contiguity between ancestor and descendant, self and family, the contemporary moment and the troubled past – an errant relationality – span Simpson and Das' works in *Oceanic Thinking*.

Two of Simpson and Das' works were titled *Vessel* (2022). One consisted of a large-scale immersive installation reframing colonial histories and perspectives on South Asian indentured labour through experimental sound. Within a makeshift housing structure of aluminium and timber, mounds of aromatic earth supported clay lota vessels emanating an arresting soundscape. The audio acknowledged the history and lived experiences of those taken from pre-Partition India to South Africa during the late 1800s and the early 1900s. The expertly crafted soundscape by Das aches, bends, and grapples with the gushing of waves over a creaky, leaky ship on the high seas, carrying the listener alongside other captive enslaved kin to the various littoral sites of indenture. Following the 1807 abolition of slavery in the British Empire, it was instead rebranded and continued as contractual indenture; this form of human trafficking, alongside the unpaid wages of First Nations peoples of Australia, created the accumulated wealth of the plantations across the northeast states of Queensland and New South Wales.

As a form that carries, both literally and here metaphorically, the clay lota vessels became ancestral objects that delivered these histories through song and enacted ritual. Against the backdrop of a stormy wavescape, the second *Vessel* work – a public art commission across the art museum's front window façade – pulled a phrase from another of Simpson's poems, 'we carry and hold these vessels' (2022, p. 54), which she transcribed on a wall inside the exhibition. The word 'vessel' is an open signifier, variably meaning body, water, boat, and by extension, museum, as a site of the ongoing colonial project. Simpson gestured to ancestral Dark Waters, both consumed and embodied, but also to submerged futures due to sea-level rise. An important vessel that Simpson grapples with is the colonial archive that both allows her to assemble understandings of her ancestors' experiences and reiterates the marked absence of their voices. Writing about undertaking research on slavery in the North American context, Christina Sharpe describes a similar lacuna:

> Those of us who teach, write, and think about slavery and its afterlives encounter myriad silences and ruptures in time, space, history, ethics, research, and method as we do our work. Again and again scholars of slavery face absences in the archives as we attempt to find 'the agents buried beneath' the accumulated erasures, projections, fabulations, and misnamings.
>
> *(2016, p. 12)*

Giving over the façade of the art museum to Simpson's words was a means by which to respond to these absences by naming them and thus affording them presence. Simpson's work reorients ownership of this history and pays homage to those forgotten, who were sent out across the Dark Waters to Natal, to the Mascarene Islands, to the Fiji Islands, to Guyana, and to the Caribbean Islands.

Seeing underwater

Izabela Pluta's sculpture *Variable depth, shallow water* 2020 (Figure 7.3) exhibited in *Oceanic Thinking (Season One)* spatialises the viewing experience, drawing on the embodied experience of diving. The sculpture – or photostructure, as Pluta refers to it – is multidirectional. A large aluminium rig comprising a central spine from which rectilinear appendages extend at 90 degrees, it requires viewers to circumnavigate its perimeter. Photographic images extend from the frame: printed aluminium plates jut sharply out, hinged mirrors and lengths of fabric are printed with maps, which drape and hang loosely towards the floor. This is a work that pulls in a number of directions, but foremost is Pluta's use of both spatial and representational means to signal a different modality of vision – seeing underwater – and a critique of representational conventions relating to marine images, especially cartography.

FIGURE 7.3 Izabela Pluta. *Variable depth, shallow water*, 2020. Installation view, *Oceanic Thinking, 2022*, UQ Art Museum.

Source: Courtesy of the artist, Newcastle and Gallery Sally Dan-Cuthbert, Sydney. Photo: Joe Ruckli.

Through its spatial formation *Variable depth, shallow water* manifests an explicitly embodied viewing experience, in which seeing the work is contingent on navigation and movement. Melody Jue's thinking in her book *Wild Blue Media* is useful for understanding this embodied viewer in relation to diving. Establishing 'thinking through seawater' as a methodology of disruption in the field of media studies, Jue positions the ocean not (only) as the object of study, but as a specific milieu through which to develop a 'conceptual displacement' that defamiliarises assumptions, structures, and meanings informed by our usually terrestrial conditions (2020, p. 163). Jue conceives of the ocean in relation to embodied epistemologies, which can dislodge normative ways of thinking habituated through our relationships to the ground, air, and gravity – relationships that are usually unrecognised. It is important to Jue's methodology that the ocean is brought into dialogue with terrestrial forms of knowledge and interpretation. In a chapter on underwater museums – both natural formations and art installations on the seabed – she deploys diving as praxis, and argues that the conditions of viewing underwater draw attention to the corporeality of the viewing experience:

> In the ocean or in a cavern, there is no Plexiglas boundary between the visitor and the historic object or formation, and the visitor is responsible for negotiating their own body movement so as not to crash into them. As a result, the visitation involves not only visual spectacle but also the embodied experience of moving around the sculptures in a fluid medium, making it difficult to contemplate from one stationary spot.
>
> *(Jue, 2020, p. 163)*

This thinking is remarkably close to writing about Minimalist art practices from the mid-twentieth century, which grappled with the way in which viewers interacted with simple, large-scale sculptures often displayed on the floor. In artist Robert Morris' terms: '[one is aware of] oneself existing in the same space as the work . . . establishing relationships as [one] apprehends the object from various positions and under varying conditions of light and spatial context' (1995, p. 232; see also Fer, 1999; Fried, 1967; Kwon, 2009). In its clean, machined lines and metal form, the aluminium structure in *Variable depth, shallow water* draws on this lineage, calling to mind Sol LeWitt's gridded sculptures. Pluta makes recourse to this legacy, drawing the viewer into a dynamic, mobile, critically embodied encounter with the work to manifest the kind of spatial negotiation Jue describes. Where the Minimalists eschewed representation, taking 'relationships out of the work and [making] them a function of space, light, and the viewer's field of vision' (Morris, 1995, p. 232), Pluta adulterates her grid with various images: underwater photographs, maps, whatever the mirror reflects – the floor, at times the viewer's lower body. Importantly, at no point is a viewer able to see all images – to see this work is to navigate it, to move and use your body to see.

The primary images in *Variable depth, shallow water* are of the Tieqa-Tad Dwejra (known in English as the Azure Window) in Gozo, an island off the coast of Malta. Rising 28 metres out of the sea, the famous arch developed from wave erosion on the rocky cliff face and was a popular tourist destination until its collapse after intense storms in 2017. Pluta's underwater images – that she captured by way of diving – of the collapsed arc are greyscale, nondescript, the iconic arc dissolved to rubble. With no *form* (no icon) discernible the representational content of the images takes a backseat to their materiality. In other words, Pluta draws attention to the way the image is made, makes us see the photograph as a photograph, before we see the piled, underwater rocks. Pluta's work is invested in effecting exactly this kind of consciousness about photography as an act – something undertaken by what Roland Barthes refers to as an 'operator' (1980/2000, p. 9) – and as a mediating form.

Other images in *Variable depth, shallow water* are sourced from the rare 1974 *Atlas Okeanov* or Atlas of the Oceans, a book of maps from the Soviet Union's War Department presenting data collected by its network of drifting stations. Unhinging and reorienting pages from the *Atlas Okeanov*, Pluta deliberately questions and complicates its record. In the darkroom she exposes the original pages to photographic processes; as enlarged light passes through the paper's surface it fuses together the maps printed on both sides, presenting them as new, unstable images, areas of which lack focus and clarity, with information redacted or otherwise muffled. As with her underwater photographs, Pluta forces a confrontation with the material conditions of maps as images, as things that are made and that can be remade, or unmade.

In making her case for the need for a 'critical ocean studies', Elizabeth DeLoughrey discusses the importance of visual forms of representation in asserting and sustaining military prowess, noting American examples in use over many decades to the present moment: '[as] a mode of warfare, the United States military's visual reproduction of its destructive power over sea and airspace – the global commons – continues today' (2019, p. 23). Despite this – and despite the expansion of transoceanic militarism – DeLoughrey asserts that outside of 'Pacific–Indigenous studies' there is a dearth of scholarship on, and literary, artistic, and cultural critique responding to, the militarisation of oceans (2019, p. 26).[6] Pluta offers exactly this kind of criticality in her treatment of the *Atlas Okeanov*. She deliberately distorts the accuracy and infallibility of what we have come to expect from mapping as a tool of surveillance, especially through settler colonial expansion into the age of the ubiquitous Google Maps. While Pluta's work is derived from her engagement with the Soviet maritime atlas and an underwater site in the Mediterranean, the critical awareness she enlivens through this work is pertinent for the waters of our highly militarised region of the Great Ocean, as discussed in the following, and final, section.

Drowning Western imaginaries

We turn now to Angela Tiatia's work *The Pearl* 2022 (Figure 7.4), exhibited as part of *Oceanic Thinking (Season Two)*. In the single channel video installation Tiatia layers complex symbols in a digitally rendered monument composed predominantly of stacked, pink plastic clamshell swimming pools, overflowing with water and creating a fountain. Tiatia was originally intrigued by the small bronze sculpture of the goddess Venus made by Henri Matisse during his 1930s travels across the islands. In Greco-Roman mythology Venus is often depicted emerging from a shell, arising from seafoam. This origin story of Venus is predated by the Great Ocean creator–god Ta'aroa, who was born from a clam shell. Surveying the aesthetics of her own Sāmoan culture and while travelling to Tahiti, Tiatia sought objects often sold as tourist items – porcelain figurines of women adorned in ula (garlands), conch shells, frangipani wreaths. She landed on the ubiquitous plastic clam shell. Durable, accessible, and affordable, it is often used by children as a pool in the yard to cool down during summer. Born from Tiatia's oceanic fountain of pink clams are orchids, birds of paradise, a whole suckling pig adorned with candy, cowrie, ula, fa'i (bananas), among other items signalling ceremony. In its impressive screenspan, *The Pearl* elevates these forms to a siapo viliata (animated or digital barkcloth) (Eshrāghi, 2021) or tivaevae (embroidered quilt), the camera panning to allow the audience to pause, and hover across and around the objects.

FIGURE 7.4 Angela Tiatia. *The Pearl*, 2021–22 (video still). Single channel digital video, colour, sound. 7 minutes, 26 seconds. Aspect ratio 32:9. edition of 8 + 2 Aps.

Source: Image courtesy of the artist, Sullivan+Strumpf, Sydney.

Bright and kitsch in aesthetic, there is a subterfuge at play in *The Pearl*, a criticality belied by its colour and rhythm. Banaban, I-Kiribati, African American scholar Teresia Teaiwa's (1999) concept of 'militourism' is key to understanding the visual language in the work, especially the layered symbolism of touristic commodities. Teaiwa coined the term to capture the intricately connected structures of colonial militarism and essentialist tourism in the Great Ocean region. Teaiwa defined militourism in multiple contexts across the region as rooted in Fernão de Magalhães' 1521 encounters with CHamoru in Guåhan and as expanded by various empires emboldened by settler colonial entitlement. Militourism, Teaiwa argues, 'is a phenomenon by which military or paramilitary force ensures the smooth running of a tourist industry, and that same tourist industry masks the military force behind it' (1999, p. 251). Teaiwa's critique pertains to most island groups but particularly to contemporary militourist presences in Rapa Nui by Chile, Guåhan, and the Hawaiian islands by the United States, and Tahiti, Pora Pora, and Grande Terre by France, among the numerous archipelagos grouped in the French colonial imaginary as French Polynesia, Wallis and Futuna, and Kanaky/New Caledonia. We understand this phenomenon as reaching to Darwin, Alice Springs, and Townsville as militourist regions of northern Australia.

In a region famously described by Epeli Hau'ofa as a 'sea of islands' (1994, p. 152), it is unsurprising that militourism manifests its impacts on the ocean and people alike, as Teaiwa makes clear: '[in] French Polynesia nuclear testing at Moruroa and Fangataufa has eroded reefs and poisoned sea and human life, belying the paradisical advertisements for Club Med at Borabora' (1999, p. 252). According to Teaiwa, the function of militourism is undergirded by the flattening region-wide imagery of a nondescript paradise. Central to that myth is the Western fantasy of the sexually available 'vahine', the monolithic Polynesian woman's body, a construct that is at once idealised, homogenised, unrealistic, misogynist, and racist, and which functions to dispel the diversity of the thousands of Indigenous peoples of multiple genders living and who have lived across the Great Ocean.

Tiatia brings an acute awareness of the hypervisibility of all things – material and ceremonial culture, environments, people – related to the Great Ocean and of all visuality pertaining to racialised bodies.[7] Tiatia, who has often included her own body in her works, had initially conceived *The Pearl* with performers in mind. Due to the pandemic, filming for the work was delayed, and ultimately she decided to take the opportunity to progress the work without live performers, effectively withholding visual access to the projected image of the 'Polynesian' body but leaning on the powerful, rhythmic score to signal the embodied practice of sāuniga (ceremony). In this way, *The Pearl* advances a critique of stereotypes associated with her community, and the broader communities of the Great Ocean, while resisting the racial hierarchies insinuated by the Western construct of the Polynesian body. Tiatia's work instead offers audiences a way out from imposed, naturalised race hierarchy, by displaying embodied practices precisely without the visually charged figures of Western imaginaries.

The Pearl holds in tension a knowing criticality regarding the colonial commodification of Great Ocean cultures and peoples, with a powerful assertion of practices and epistemologies that dwarf such exploitations. At the end of the iteration of *The Pearl* shown in *Oceanic Thinking (Season Two)*, the camera pulls back from the bright formation of the clam shell fountain and pans down cascading water to reveal a triangular malae (central ceremonial clearing in Sāmoan villages) rising out of the ocean. On top of the malae sits a penina (pearl) sliced through with a sapelu (machete) as deity-sized women, digitally modelled after mass-produced figurines, watch omnipresent from the surrounding waters. Largely submerged, their visibility is withheld, reclaiming them from the militourist economy and returning them to the ocean. Mangioni's assertion that people from the Great Ocean are instilled with an 'environmental literacy of ourselves in relation to our surroundings' (2019) is in evidence here. Indigenous conceptions of the stewardship of territory in Sāmoa, Tonga, Viti (Fiji), Hawai'i, among other archipelagos, render reciprocity and interdependence with the waters, plants, animals, and bird kin alike, from the ridge to the reef, and from the nine heavens to the multiple underworlds. As Mangioni puts it: 'Our place in and by the water' (2019).

A return to the ocean

Across their diverse practices Simpson, Pluta, and Tiatia all engage with the blue. Individually they are motivated by specific and inherited modalities of knowledge, with an attentiveness to the material, historical, political, and cultural potencies of blue perspectives, each demonstrating the generative potentiality of thinking about, through, and nearby water. The *Oceanic Thinking* exhibitions were informed by a current interest in this kind of practice, and a desire to respond in kind, deploying what Hessler calls 'tidalectic curating'. An accountability for the politics, and carbon emissions, of exhibiting institutions is one part of this work, and our carbon neutral certification for the first season of *Oceanic Thinking* was an important way in which climate change was reckoned with through the exhibition. In this chapter we have focused on artists that render the oceans as cultural and historical sites, as a milieu (to use Jue's term) through which modalities of knowledge emerge that disrupt certain terrestrial assumptions. The governing desire of the *Ocean Thinking* exhibition was to tell multiple, diverse stories about oceans and their cultures, gathering voices for future *Blue Assembly* projects to develop and further investigate terrestrial sites demarcated or haunted by water.

Through this discussion of a number of key works by Australian-based artists living and working with and in many blue places, we made manifest our commitment to approaching a tidalectic curatorial practice in the interdisciplinary and pedagogical context of the University of Queensland Art Museum through relational ethics that refocus again and again on increasing literacy in the Majority World. The screens, photostructures, soil and lota vessels, glass windows, and paper with which Sancintya Mohini Simpson, Isha Das, Izabela Pluta, and Angela Tiatia work, enable us to consider relationships and responsibilities outside of empire and capital.

By employing blue 'research-creation' they bring together ways of knowing and being that, while aligned with the humanities, also draw on Indigenous or marginalised knowledge systems, all of which model and enact the kind of collectivity and collaborative solidarity urgently needed to secure futures for all species and spaces on this planet. The blue turn in contemporary art is evidenced then, by the relational and aestheticised spaces between humans and more-than-human species and spaces. Further, the blue turn is characterised by the ways in which artists, viewers, and artworks are submerged symbolically and physically in aqueous states that signal a desire to belong to, and be part of, future-making through, with, and nearby the Majority World.

Notes

1 According to Truman and Springgay, 'Research-creation can be described as the complex intersection of art practice, theoretical concepts, and research. It is an experimental practice that cannot be predicted or determined in advance. It is transdisciplinary and is used by artists and designers who incorporate a hybrid form of artistic practice between the arts and science, or social science research; scholars attuned to the role of the arts and creativity in their own areas of expertise; and educators interested in developing curriculum and pedagogy grounded in cultural production. Research-creation is attuned to processes rather than the communication of outputs or products' (Springgay and Truman, 2018).

2 This distance is iterated by critic Erik Morse, who states: 'Critical ocean studies' waterborne imaginaries . . . have ventricles that stretch back far beyond a twenty-first-century eco-poetics. In fact, much of the current artistic fascination with these imaginaries is indebted to a premodern worldview, in which climate was often associated with a sublime, and leaky, volatility' (2022, p. 258).

3 The exhibition Oceanic Thinking was also awarded the Queensland Museums and Galleries Award for Sustainability for achieving carbon neutrality certified by Climate Active, the Australian Federal Government's Clean Energy Regulator.

4 While the term 'Majority World' in more recent parlance refers to racialised peoples of this planet in opposition to the colourist racialising binary of White people at the centre and Black, Indigenous, and other Peoples of Colour at the periphery, we expand its use to directly address and consider ourselves alongside more-than-human species and spaces that are the majority of living entities and ecologies: that is to say, humanity is the minority among many.

5 Blackbirding refers to the practice of kidnapping Indigenous peoples of the archipelagos today grouped as West Papua, Papua New Guinea, Solomon Islands, Vanuatu, Kanaky/New Caledonia, Tokelau, and Niue among others into forced labour on sugarcane and cotton plantations along the eastern Australian seaboard.

6 We understand DeLoughrey to be referring predominantly to Euro-American intellectual and cultural production.

7 In 2022, Tiatia and Eshrāghi's public conversation at University of Queensland Art Museum was titled 'Hydrodecadence', a term that Eshrāghi has put forward to capture the tension between, on one hand, this hypervisibility of Oceanic cultures and people and, on the other, the exploitation of natural water resources on Indigenous territories. They point to the example of the water supply from Quebec dams on Cree and Inuit homelands which is exclusively available free of charge to larger urban areas and not Indigenous communities on ancestral territories.

References

Barthes, R. (2000). *Camera Lucida: Reflections on Photography*. London: Vintage. (Original work published 1980).

Day, C., & Ratcliffe, M. (Eds.). (2021). *Tree Story*. Melbourne: Monash University Museum of Art and Monash University Publishing.

Deleuze, G., & Guattari, F. (2004). *A Thousand Plateaus: Capitalism and Schizophrenia*. London: Continuum.

DeLoughrey, E. (2019, April). Toward a critical ocean studies for the Anthropocene. *English Language Notes*, *57*(1), 21–36.
Eshrāghi, L. (2021). *Sāmoan Art Museum-Performance*. Toronto: A Space Gallery and imagineNATIVE.
Fer, B. (1999). Objects beyond objecthood. *Oxford Art Journal*, *22*(2), 27–36.
Fried, M. (1967). *Art and objecthood.* Retrieved 17 June 2018 from https://monoskop.org/images/8/8f/Fried_Michael_Art_and_Objecthood_Essays_and_Reviews_1998.pdf
Glissant, É. (1997). *Poetics of Relations*. The University of Michigan Press. (Original work published 1990).
Haraway, D. (2015, May). Anthropocene, capitalocene, plantationocene, chthulucene: Making kin. *Environmental Humanities*, *6*(1), 160.
Haraway, D. (2016, September). Tentacular thinking: Anthropocene, capitalocene, chthulucene. *e-flux* (#75). Retrieved 10 December 2022 from https://www.e-flux.com/journal/75/67125/tentacular-thinking-anthropocene-capitalocene-chthulucene/
Hau'ofa, E. (Spring 1994). Our sea of islands. *The Contemporary Pacific*, *6*(1), 147–161.
Havini, T., & Taumoepeau, L. (2020, October). *The Last Resort: A Conversation. e-flux* no. 112. Retrieved 10 December 2022 from https://www.e-flux.com/journal/112/353919/the-last-resort-a-conversation/
Hessler, S. (2020). Tidalectic curating. *Journal of Curatorial Studies: Curating the Sea Special, Issue 9*(2, October), 248–270.
Iovino, S., & Oppermann, S. (2014). *Material Ecocriticism*. Bloomington: Indiana University Press. muse.jhu.edu/book/41333
Jue, M. (2020). *Wild Blue Media: Thinking Through Seawater*. Duke University Press.
Kwon, M. (2009). Approaching architecture: The cases of Richard Serra and Michael Asher. *Yale University Art Gallery Bulletin State of the Art: Contemporary Sculpture*, 44–55.
Mangioni, T. L. (2019, September). Oceanic sovereignty, climate justice and art-story. *Runway Journal: Oceans*, *39*. Retrieved 16 December 2022 from http://runway.org.au/oceanic-sovereignty-climate-justice-and-art-story
Moore, J. W. (2017). The capitalocene, part I: On the nature and origins of our ecological crisis. *The Journal of Peasant Studies*, *44*(3), 594–630. doi: 10.1080/03066150.2016.1235036
Moreton-Robinson, A. (2015). *The White Possessive: Property, Power, and Indigenous Sovereignty*. University of Minnesota Press.
Morris, R. (1995). Notes on sculpture: Parts one and two. In G. Battock (Ed.), *Minimal Art: A Critical Anthology* (pp. 222–235). University of California Press.
Morse, E. (2022, June). Why artists are returning to 'oceanic thinking'. *ArtReview*. Retrieved 16 December 2022 from https://artreview.com/artists-oceanic-thinking-consciousness-water-aquatic-turn/
Sharpe, C. (2016). *In the Wake: On Blackness and Being*. Durham: Duke University Press.
Simpson, S. M. (2022). Gone to Tāpū. In O. Fairweather, L. Nowell, H. U. Obrist, A. Raza, & W. Weatherall (Eds.), *This Language that is Every Stone: Édouard Glissant and the Australian Archipelago* (pp. 54–55). Institute of Modern Art; Verlag der Buchhandlung Walther und Franz König. https://www.ima.org.au/product/this-language-that-is-every-stone/
Simpson, S. M., & Das, I. (2022). Email correspondence, 10 February 2022.
Springgay, S., & Truman, S. E. (2018). *Walking Methodologies in a More-than-human World: Walkinglab* (First edition). Taylor and Francis.
Teaiwa, T. (1999). Reading Paul Gauguin's *Noa Noa* with Epeli Hau'ofa's *Kisses in the Nederends*: Militourism, feminism and the 'Polynesian' body. In V. Hereniko & R. Wilson (Eds.), *Inside Out: Literature, Cultural Politics, and Identity in the New Pacific* (pp. 249–263). Rowman & Littlefield.
Tinker, H. (1988). New lights from old languages: Indian languages and the experience of indentureship in South Africa. In S. Bhana (Ed.), *Essays on Indentured Indians in Natal* (p. 198). Peepal Tree Press.

PART 3
Mediating the blue

8

ECOPOLITICS AND ECOLOGICAL IMPERIALISM

Activists, artisans, and the Save the Reef campaign

Maxine Newlands

In the face of escalating threats from anthropogenic climate change, the urgency to comprehend its multifaceted dimensions has never been more pressing. While scientific and policy-oriented perspectives dominate discussions, a compelling need exists to integrate the humanities and social sciences into our understanding of climate change impacts. This chapter explores the significance of the blue humanities through a study of Australia's Great Barrier Reef, to spotlight the intricate relationship between human societies and the world's oceans, emphasising the profound impact of climate change on our maritime environment.

The blue humanities encourages us to look beyond the surface, recognising the vital role of water systems, such as the ocean, as more than vast expanses of water to be exploited and traversed but as dynamic systems integral to how nature regulates the climate and related ecosystems. With the ocean making up 70 per cent of the Earth's surface, it plays a pivotal role in absorbing and storing carbon dioxide, influencing weather patterns, regulating global temperatures, and creating a mass of biodiversity. Through the blue humanities, we can reach a deeper understanding of the interconnections between the oceans and climate, gaining a more nuanced cultural and social perspective on the consequences of climate change.

Climate change is not merely an environmental phenomenon; it is an anthropogenic-induced problem, a wicked problem that requires different ways of understanding and interpreting the biophysical sciences. Whilst biophysical science is commonly presented in a positivist way, there are alternative ways of translating scientific data into varying platforms. The blue humanities provides a platform to amplify the voices of those directly affected, weaving scientific data into narratives to convey the emotional and cultural ties humans have with the blue. These stories anthropomorphise the climate crisis and provide a better

DOI: 10.4324/9781003365501-12

public understanding of the consequences and impacts of climate change on our oceans and water systems.

The blue humanities enables us to draw upon literature, art, and historical records to reconstruct past climates and understand how societies coped with environmental changes. For example, the early work from Rachel Carson, *The Sea Around Us* (1951), mixes Carson's scientific expertise in early marine science with historical facts, poetry, artefacts, policy design, and scientific data. More recent works, such as Anna Elias' book *Coral Empire – Underwater Oceans, Colonial Tropics, Visual Modernity* (2019), show us how, through arts, photography, and filmmaking, the humanities can better understand the cultural significance of corals, fisheries, and landscapes and how our contemporary knowledge is defined by colonial thinking. Ian McCalman's *The Reef: A Passionate History* (2013) tells us of how humans attempt to conquer and exploit the Great Barrier Reef through historical facts and biographical facts, and the documentary *Chasing Coral* (Coomb et al., 2017) combines interviews with marine scientists with footage of bleaching corals in varying states of death or dying ecosystems – at the same time, the footage and narrative go some way towards 'coral reef is among the most photographed, most fetishized, and most exploited of all oceanic phenomena' (Elias, 2019, p. 234). These examples show how the blue humanities compels us to confront ethical questions surrounding our responsibility as custodians of the oceans.

Climate change is a direct threat to marine biodiversity and the cultural heritage tied to it. Coral reefs, marine ecosystems, and traditional practices are at risk. The blue humanities underscores the importance of preserving biophysical, social, and cultural knowledge by recognising how the interconnectedness of cultural and ecological systems should foster a holistic approach to climate mitigation and adaptation. Understanding our responsibility to shape these environments is essential as climate change impacts accelerate. Examining maritime ethics through the blue humanities lens, combined with acceptance to change our ecological imperialistic languages and framing, prompts a critical evaluation of our relationship with the oceans and the unethical actions that continue to undermine the resilience of our oceans and waterways.

This chapter presents two case studies that highlight how cross-disciplinary research between the natural sciences, humanities, and social sciences can be effective, and echoes Reef historian Professor Iain McCalman's call for more collaborations between the biophysical and social sciences and the humanities, to share knowledge systems and seek out transformative change. Indeed, a point reinforced by McCalman, who notes that

> the overwhelming majority of interviewers and reviewers presumed automatically that I must be a marine scientist, and tailored their questions accordingly. This held true even after I explained that I was a historian and attempted to focus their attention also on the aesthetic, psychic, cultural and social dimensions of the Reef. Australian scientists cannot be blamed for this exclusivity of focus.
>
> *(2017, p. 72)*

Instead, McCalman proposes that Reef scientists need to work with environmental humanities and arts practitioners to reach

> crucial potential allies in the tourist industry who depend on the beauty, health and wonder of Reef corals and their waters to attract thousands of national and international visitors each year to enjoy the aesthetic and sensual pleasures entailed in recreational swimming, fishing and coral viewing.
>
> *(2017, p. 74)*

This chapter precisely reinforces this proposal within a blue humanities framework to show where there is potential to facilitate connections across the disciplines, changing the source of knowledge systems beyond the biophysical sciences to include transdisciplinary scholarly literature, practitioners, Indigenous Knowledge, and local community groups. Viewing the present-day dilemmas that come with anthropogenic climate change impacts through a blue humanities lens allows us to discern patterns, shifts, and adaptations.

The first case study is one of the world's first environmental campaigns, Save the Barrier Reef (1966–70), to show how interconnectivity between the arts (humanities), politics (social science), and marine science (natural science) can be more effective than individual disciplines. This campaign, which began in 1966/67 (Lloyd, 2022; McCalman, 2014), is useful in identifying how the numerous scholars, practitioners, scientists, local community members, and politicians were able to collaborate and what this might mean for future understanding of the complexities of human–environmental relationships today, and identifying new modes of knowledge necessary to guide global decision-makers.

The second case study is the Reef Restoration and Adaptation Program (RRAP), a transdisciplinary project with marine scientists, environmental social scientists, economists, decision-support modellers, mathematicians, Aboriginal and Torres Strait Islander peoples, policy analysts, federal government regulators, citizen scientists, and others. The Program is a universal joint venture between Australia's leading marine science universities, scientific institutions, and government reef protection agencies. RRAP's aim was to show there were innovative and novel scientific techniques that can be scaled-up to help coral reefs in the Great Barrier Reef and other biodiversity systems become more resilient against warming waters and bleaching episodes.

As this chapter will show, successful outcomes are stronger when bringing together scholars, politicians, advocacy groups, and local community groups by forming a symbiotic relationship on a unifying problem.

What is the Great Barrier Reef?

The Great Barrier Reef is a collection of corals, reefs, clays, and islands, home to hundreds of species. Geographically larger than the country of Italy, the Reef traverses through many latitudes, from the far north of the Torres Strait Islands to the southern point near Bundaberg. The Reef ebbs and flows along the Queensland coastline, as the once rainforest land submerged by sea is the Reef of today. The northern part of the Reef hugs the shoreline, moving away from the land as the wet tropics turn into the dry tropics, and the Reef moves into the sub-tropics below the tropic of Capricorn.

Case Study One – saving the Great Barrier Reef

Scientists and environmentalists, alarmed by the escalating pressures from plans to increase agro-based products, mainly sugar cane, and a realisation of the potential harm capitalist ventures could do to the environment, joined forces to raise awareness about the potentially precarious state of the Great Barrier Reef. The trigger was an application to mine Ellison Reef in Far North Queensland (1967) for limestone because the Reef was dead. Limestone can balance the Ph in the soil to produce higher-quality sugar cane (Lloyd et al., 2017). A group of local artists and activists began a movement that was to become a multidecadal ideological clash between the blue and the politics of economic growth, mainly from the mining and agricultural industries and some of the tourism sector.

The first Save the Barrier Reef campaign, 1967

It is 1967, and John and Alison Büsst sit on the deck of their home surrounded by the sounds from the tropical rainforest. Geographically located between the Great Barrier Reef and tropical rainforest of Far North Queensland, with the swaying palm trees, lifted by the treble of bird songs, grounded by the scuttling of small animals hunting food, all accompanied by the melodic ocean waves, breathing in and out, in and out, in and out on the shoreline below, this was a home made from bamboo perched above the coastline of Bingil Bay.

Büsst shared a history and relationship with the famous beachcomber and author Edmund Banfield, who owned a property on nearby Dunk Island. Dunk Island and its sister island Bedarra sit adjacent to Bingil Bay in the Great Barrier Reef marine park. Bedarra was purchased by Captain Henry Allason from the Queensland Lands Department in 1913. Forced to return to Europe with the outbreak of the First World War, Allason sold the property to the Harris syndicate of London in 1934. In the late 1930s, famous Australian landscape artist Noel Wood purchased East Bedarra and over time, it was developed into a luxury tourist destination. By the mid-1940s, Büsst and his sister Phyllis had purchased part of the island.

Following several cyclones, the Büssts moved to the mainland and built their home in the rainforest, Ninney Rise. Ninney, as it is affectionately known, became the headquarters of the first Save the Barrier Reef campaign in 1967. Along with Australian poet Judith Wright and student marine scientist Eddie Hegerl, John Büsst was also an old school friend of Australia's then Prime Minister Harold

Holt. After university, Büsst lived and worked at the Montsalvat artists' colony in Eltham, a suburb of Melbourne. Former pupil Sigmund Jörgensen told the former President of Friends of Ninney Rise, Tony O'Malley,

> After he left uni, John took up with painting with my father and helping building. He was here at Montsalvat about 1935 to 1944 when he left and bought Bedarra Island. He had romantic illusions to build an artists' community there. He was very heavily involved in the conservation of the Barrier Reef. Harold Holt in a speech thanked Büsst for his contribution to reef conservation. The Holts built a house close by and they visited each other a lot.
>
> *(personal communication, July 2014)*

In 1967, a coalition of scientists, artisans, a poet, and a prime minister, along with the community and environmental and national conservation groups, formed an alliance to prevent the mining application and formed the Save the Barrier Reef campaign. This grassroots movement aimed to bring the urgent need for reef preservation to the forefront of public consciousness. Recognising the importance of public engagement, the campaign adopted a multi-pronged approach that combined scientific research, advocacy, and public outreach. Their research provided crucial insights into the Great Barrier Reef's ecological threats, emphasising the need for immediate action.

Whilst Holt privately backed the campaign until his disappearance in December 1967, as President of the Innisfail branch of the Wildlife Preservation Society of Queensland (WPSQ) Büsst could rally environmentalists across Queensland and Australia to great effect. The Save the Barrier Reef campaign recruited Hegerl and his fellow student marine scientists from the Queensland Littoral Society to survey the Reef. The Wildlife Preservation Society of Queensland (WPSQ) held barbeques across the state to raise funds and organise letter-writing campaigns to bombard politicians and journalists into taking up the cause. Every letter contained a sticker or bumper sticker – 'Save the Barrier Reef' – and soon they were appearing across Australia. Armed with the scientific proof from Hegerl's survey that Ellison Reef was alive, Büsst was successful in challenging the mining application and, in the process, immobilising future mining applications and ultimately scuppering plans to mine the Reef for limestone or other resources in the future (Lloyd, 2022).

The campaign of 1967 left an enduring legacy, and not only contributed to the immediate protection of the Great Barrier Reef but also set a precedent for future environmental movements.

The WPSQ, beneficiaries of Phyllis Büsst who donated money to the WPSQ, and close friend of Australian poet and environmental and Indigenous rights advocate Judith Wright, had a significant role in coordinating the Save the Reef campaign, by bringing Queensland and international conservation and environmental groups together.

Politically, the Save the Barrier Reef campaign led to a national recognition of the Reef systems as the world's largest ecosystem – the Great Barrier Reef. Australia's government passed the Great Barrier Reef Marine Park Act (1975), making it the world's first marine national park, with the United Nations Educational, Scientific, and Cultural Organization's (UNESCO) granting of World Heritage Status in 1981 seemingly securing the future health of the Barrier Reef.

Whilst this story might be one of artists, authors, and advocates continuing the conservation stories of Banfield and Büsst, it was also heavily influenced by the political and social networks between the main protagonists. John Büsst could see that if one mining application could be granted, there would be no stopping further applications the length of the Great Barrier Reef.

A photograph taken at Ninney Rise during a dinner party gives clues as to another life of Büsst, one of power brokering, influence, and networks.

The Büssts and Australian poet and conservationist Judith Wright became acquainted when rallying support against the mining application. Büsst wrote to her, requesting assistance with a 'vital test case' (McCalman, 2014, p. 276) to stop the

FIGURE 8.1 Dinner at Ninney Rise, John Büsst (centre) stands next to Zara Holt, wife of Australian Prime Minister Harold Holt to his right, Alison Büsst third from the right, and poet and author Judith Wright third from the left.

Source: Used with permission from Friends of Ninney Rise.

nearby Ellison Reef from being mined for limestone. Wright, then President of the Brisbane-based WPSQ, was happy to help, and would become the foremost author on the first Save the Barrier Reef campaign with her book *Coral Battleground* (1996).

The Save the Barrier Reef campaign of 1967 left an enduring legacy. It not only contributed to the immediate protection of the Great Barrier Reef, but also set a precedent for future environmental movements. The campaign highlighted the power of collective action, and the positive effects of combining scientific knowledge with humanities and social science research, unified through the blue.

The success of the campaign also underscored the importance of ongoing vigilance and adaptive management in preserving delicate ecosystems. In the face of climate change, the lessons learned from the Save the Barrier Reef campaign continue to resonate, inspiring new generations to safeguard our planet's natural wonders for the benefit of all living beings.

However, what was seriously omitted from the campaign was the voices and knowledge systems of the Aboriginal community near Ninney Rise, and the nearby township of Mission Beach, the main location of the campaign.

Navigating the seas of ecological imperialism

Ninney Rise in on the traditional lands of the Djiru and the Bingil Bay Aboriginal people, part of the larger Indigenous community of the Girramay people. The Djiru and Bingil Bay peoples have faced significant challenges due to colonialism, including displacement and removal from their lands. Djiru peoples and the Girramay connection to the Reef is one dominated by colonialism and ecological imperialism. In Australia, the Aboriginal and Torres Strait Islander peoples' Indigenous Knowledge system of the environment is much older than settler knowledge, estimated at more than 65,000 years old. As Elias explains, the 'global, mobile, white explorers and adventurers' who [then] benefited from knowledge of the sea from maritime peoples, especially skilled divers known ubiquitously by the colonial term "native divers" . . . treated as a servant class employed to further European knowledge' (2019).

Ecological imperialism, the blue and Indigenous Knowledges

To understand how and why colonial thinking influences our relationship with the blue, we can apply the Western concept of ecological imperialism (Gibbs et al., 2021), but first we must understand imperialism. Put simply, 'imperialism is the system by which a dominant power is able to control the trade, investment, labour, and natural resources of other peoples' (Frame, 2016, p. 88). Steinberg traces the etymology of ocean imperialism back to the Romans and their domination of the Mediterranean Sea, noting that there is a difference between *imperium* and *dominium* – where the former has the right to command, the latter has the 'power to own, use, enjoy and dispose of property' (Steinberg, 2001, pp. 64–65). It was not until the emergence of the merchant capitalists – or mercantilists – that the 'sea shifted from being a space of mysterious danger to being a space without nature, unpossessable but also unremarkable' (Steinberg, 2001, p. 105), and the 'ocean–space construction (and scientific outlook) that was to predominate during the following centuries of industrial capitalism' began. Our oceans and seas were less feared and more understood as a space to be conquered and colonised (Elias, 2019).

The relationship between ecological imperialism and water is a complex interplay that reveals the far-reaching consequences of human interventions in the natural world. Ecological imperialism as a conceptual framework is the appropriation of imperialism onto ecological systems, and refers to the historical pattern where powerful nations exploit and reshape the environments of less powerful regions for their own economic or political gain. Water, as a resource, has often been at the centre of such imperialistic endeavours. The control and manipulation of water bodies, such as rivers, lakes, and oceans, have been integral to imperialistic strategies throughout history.

The blue humanities prompts us to confront the legacies of ecological imperialism. By examining historical narratives through the lens of water, it encourages a nuanced understanding of the impact of imperialistic actions on marine ecosystems and coastal communities. It is vital that we recognise how the historical legacies of ecological imperialism influence our modern-day understanding of the oceans, and how much colonial language influences our scientific and human discourse between water and adaptation to climate change impacts is crucial to be able to reject and move forward. Thus, the blue humanities, by fostering a holistic perspective, offers a path towards reconciliation, sustainability, and a more harmonious relationship with the world's water resources. It challenges us to learn from history and chart a course towards a more equitable and respectful coexistence with the watery realms that shape our planet.

Case Study Two – Coral reef restoration and adaptation

The intervening years since the first Save the Barrier Reef campaign have witnessed herculean efforts to save the Reef from the anthropogenic impacts of climate change. Technological developments, multilayered politics from the local to global, slow adaptation of Indigenous Knowledge, and economic entanglement with environmental tragedies provide a narrative that is complex, yet retains the simple aims of the first campaign – save the Reef.

More than half a century later, thousands of marine scientists, geologists, meteorologists, engineers, and citizen scientists have surveyed and scanned most of the 2,300 kilometres of the Reef. Today, scientists are looking to find unique technological solutions to saving the Barrier Reef, political solutions through global environmental programmes, and economic solutions through blue carbon, blue economies, and the blue humanities (DeLoughrey et al., 2015; Oppermann, 2019).

Since 1998, coral bleaching episodes and awareness of the impact of warming waters on the Reef ecosystem led marine science organisations to come to realise that drastic measures were required if the Reef was to build any resilience in what the modelling showed as a year-on-year increase in global temperatures (Anthony et al., 2020). What emerged was a 'mission' to save the corals through novel and innovative science that could, in some cases, replicate the natural world and, in others, change the chemical make-up of corals, clouds, and other parts of the ecosystem in order to help make the corals more resilient.

Coral reef restoration science was not new, but at scale was a new development, and a first for Australia and the world. Climate change was pushing ocean temperatures beyond those that species can tolerate, as early as the 1980s. The changes in ocean chemistry were making the seas more acidic, and the ecosystem far less resilient (Kolbert, 2021). However, in 2014 a marine heatwave began that would last three years and include the first back-to-back bleaching episodes that affected almost 90 per cent of the Great Barrier Reef (Hughes et al., 2018).

A potential large-scale solution to assist the system in building tolerance and resistance to warming waters was first mooted by the Australian Institute of Marine Science's CEO John Gunn, and became known as the RRAP[1]. RRAP is a universal joint venture between Australia's leading marine science institutions, the social sciences, and humanities, with experts' contribution from Australia's government reef protection agencies.

RRAP is a large project, which at times has includes up to 300 scientists and experts working towards a common goal – it has been developing new scientific techniques for cooling and shading, coral aquaculture, cryopreservation, ecological intelligence, enhanced corals and treatments, modelling and decision support, moving corals, regulation, rubble stabilisation, stakeholder and Traditional Owner engagement, and translation to deployment.

There is opposition to the mission, with a small number of critiques vocal on the merits of coral reef restoration at scale. Morrison et al. argue that any improvements in the long-term future of local coral Reef restoration will require policy makers, philanthropists, scientists, other stakeholders, and the public [to] all hold a stake in restoration science and its practice (Morrison et al, 2019), whilst others argue that marine ecosystem restoration is expensive, has a high risk of failure, and distracts the conversation away from tackling global climate change (Hughes et al., 2023). It is clear that restoration will only be effective if the indirect and direct factors – drivers of decline – are addressed. However, with scale comes increasing numbers of stakeholders, cultures, and social norms to be included in the research and development of coral reef restoration (Fidelman et al., 2019). With scale also comes an opportunity to ensure no repetition of ecological imperialism as the dominant approach, but rather an inclusive approach that is consensus driven in its deployment.

There are two subprograms within RRAP that address the knowledge gap and interlocutor of Indigenous Knowledge and decision-making. RRAP has a dedicated stakeholder and Traditional Owner engagement program, which supports the development and deployment of interventions that are socially and culturally responsible, and accepted as legitimate by stakeholders, rights-holders, Reef managers, and the public. The project involves working with scientists and Traditional Owners to gain permissions to go on-Country and conduct field research, cultural understanding and competence, and co-design of research and development programs. The translation to deployment subprogram is a project that involves engagement and building capacity with industry, community, and Traditional Owners in the challenge of building a reef restoration sector that can deploy the methods and approaches developed in RRAP. There are many other projects which are co-designed in conjunction with Traditional Owners, scientists, and social scientists, who, like the predecessors of 1967, are working towards a common goal.

Conclusion

This chapter has shown how transdisciplinary collaborations between the trifactor of the biophysical science, the blue humanities disciplines of social science and environmental humanities, and Indigenous Knowledge are not new, and neither are calls for greater interdisciplinary collaborations from the environmental humanities (Holm et al., 2015; McCalman, 2017) to develop new knowledge systems. At the same time, recognising that in advocating the need for systemic collaboration between the human, social, and natural sciences there are many institutions and organisations, such as international scientific organisations, that have issued calls for the inclusion of the humanities in programmes of global environmental research (Holm et al., 2015). However, recognising the need for more interdisciplinary ideas and integration of knowledge systems can be problematic, as when competing with 'an open system with a free market, weak global politics, cultural distrust, and imperfect communications, defectors are likely to get away with cheating' (Holm et al., 2015, p. 981). Long-term interdisciplinary collaboration, although not unheard of, can be structurally and culturally difficult across faculty divides, and happens infrequently (Holm et al., 2015).

Just as the role of sciences is to observe and analyse change, systemic barriers often mean science as a knowledge system is not organised, governed, or structured sufficiently to be able to create social policy or regulation, or to heavily influence human change. Proponents of interdisciplinary research across the science and technology sectors tend to relegate human and social science research to an auxiliary, advisory, and non-essential status. Further, Holm et al. (2015) suggest this is compounded by mistrust when humanities, social science, and scientific analyses are integrated to address societal problems, such as adaptation to anthropogenic climate change impacts.

An alternative is a higher recognition that each of these fields has limitations, but together they form a stronger connection between knowledge systems. Whilst science is able to monitor, measure, and to some extent predict the biogeophysics of global change, it separates itself from investigating the main drivers of ecosystem change – the human factor (Holm et al., 2015). The blue humanities can intervene by emphasising and working on these and other barriers, and by engaging closely with the natural sciences.

This chapter makes a case for what the blue humanities can do in shifting Australian environmental history from the land to the sea. Incorporating the blue humanities into our discourse on climate change can enrich our understanding of the complex, symbiotic relationship between humans and the blue to not only deepen our appreciation for the oceans, but also to provide a vital framework for shaping future pathways for a sustainable and resilient future for our planet.

Note

1 The author was a co-lead on the RRAP Regulatory subprogram, and researcher to the Engagement subprogram between 2017 and 2023.

References

Anthony, K. R., Helmstedt, K. J., Bay, L. K., Fidelman, P., Hussey, K. E., Lundgren, P., Mead, D., McLeod, I. M., Mumby, P. J., Newlands, M., Schaffelke, B., Wilson, K. A., & Hardisty, P. E. (2020). Interventions to help coral reefs under global change – a complex decision challenge. *PLOS One*, *15*(8), e0236399. https://doi.org/10.1371/journal.pone.0236399

Coomb, D., Curtis, V., & Orlowski-Yang, J. (2017). *Chasing Coral* [Netflix].

DeLoughrey, E., Didur, J., & Carrigan, A. (2015). *Global Ecologies and the Environmental Humanities: Postcolonial Approaches*. Routledge.

Elias, A. (2019). *Coral Empire: Underwater Oceans, Colonial Tropics, Visual Modernity*. https://doi.org/10.1215/9781478004462-003

Fidelman, P., McGrath, C., Newlands, M., Dobbs, K., Jago, B., & Hussey, K. (2019). Regulatory implications of coral reef restoration and adaptation under a changing climate. *Environmental Science and Policy*. https://doi.org/10.1016/j.envsci.2019.04.016

Frame, M. L. (2016). The neoliberalization of (African) nature as the current phase of ecological imperialism. *Capitalism, Nature, Socialism*, *27*(1), 87–105. https://doi.org/10.1080/10455752.2015.1135973

Gibbs, M. T., Gibbs, B. L., Newlands, M., & Ivey, J. (2021). Scaling up the global reef restoration activity: Avoiding ecological imperialism and ongoing colonialism. *PLOS One*, *16*(5), e0250870. https://doi.org/10.1371/journal.pone.0250870

Holm, P., Adamson, J., Huang, H., Kirdan, L., Kitch, S., McCalman, I., Ogude, J., Ronan, M., Scott, D., Thompson, K., Travis, C., & Wehner, K. (2015). Humanities for the environment – a manifesto for research and action. *Humanities (Basel)*, *4*(4), 977–992. https://doi.org/10.3390/h4040977

Hughes, T. P., Baird, A. H., Morrison, T. H., & Torda, G. (2023). *Principles for Coral Reef Restoration in the Anthropocene*. One Earth.

Hughes, T. P., Kerry, J. T., & Simpson, T. (2018). Large-scale bleaching of corals on the Great Barrier Reef. Ecology (Durham), 99(2), 501–501. https://doi.org/10.1002/ecy.2092

Hundlow, T. (1985). The environment. In A. Patience (Ed.), *The Bjelke–Petersen Premiership 1968–1983: Issues in Public Policy*. Longman Cheshire.

Kolbert, E. (2021). *Under a White Sky: The Nature of the Future*. Crown Publishers.

Lloyd, R. (2022). *Saving the Reef: The Human Story Behind One of Australia's Greatest Environmental Treasures*. University of Queensland Press.

Lloyd, R., Newlands, M., & Petray, T. (2017). Coral battleground? Re-examining the 'Save the Reef' campaign in 1960s Australia. *Environmental Sociology*, *3*(1), 54–63. https://doi.org/10.1080/23251042.2016.1259604

McCalman, I. (2014). *The Reef: A Passionate History* (First American edition). Scientific American: Farrar, Straus and Giroux.

McCalman, I. (2017). Linking the local and the global. What today's environmental humanities movement can learn from their predecessor's successful leadership of the 1965–1975 war to save the Great Barrier Reef. https://doi.org/10.3390/h6040077

Morrison, T. H., Adger, W. N., Brown, K., Lemos, M. C., Huitema, D., Phelps, J., Evans, L., Cohen, P., Song, A. M., Turner, R., Quinn, T., & Hughes, T. P. (2019). The black box of power in polycentric environmental governance. *Global Environmental Change*, *57*, 101934. https://doi.org/10.1016/j.gloenvcha.2019.101934

Oppermann, S. (2019). Storied seas and living metaphors in the blue humanities. *Configurations (Baltimore, Md.)*, *27*(4), 443–461. https://doi.org/10.1353/con.2019.0030

Steinberg, P. E. (2001). *The Social Construction of the Ocean.* Cambridge University Press.

Wright, J. 1915–2000. (1996). *The Coral Battleground / Judith Wright.* Pymble, N.S.W.: Angus & Robertson.

9

DIGITAL BLUES

Sense of self and the human–nature–technology connection in Australian aquatic environments

Melusine Martin

Digital technology is changing the way we relate to blue spaces, allowing us to experience natural environments online beyond the limits of geographical boundaries and time zones. Living in a mixed reality – at the intersection of reality and virtuality – impacts the way we relate to our body, to our self, and to place. Digital nature, which refers to visual, auditive, and textual components related to natural environments experienced via digital technology, can raise our awareness of aquatic environments, notably through debates on climate change and plastic pollution. But it can also affect the human perception of and relationship with blue environments by generating a mind–body disconnect. While digital technology can expand our sense of self to remote natural environments, it can also trigger what this chapter defines as digital solastalgia – the distress resulting from the online experience of environmental change and the damage done to our extended environment. In this chapter, I investigate how Australian pro-environmental digital users connect to aquatic nature in a digital context. The research design consisted of an online survey of over one hundred pro-environmental digital users followed by semi-structured, in-depth interviews with ten interviewees.[1] Blue spaces – rainforest creeks, waterfalls, estuaries, reefs, and, of course, the ocean – were an important aspect of the participants' bond to nature, as most of them lived in coastal areas of Australia. But, as daily active digital users, they partook in an improbable, paradoxical routine, which has quickly become the norm in Western countries: they simultaneously lived in two worlds – real and cyber. They could both experience the blue environments of their immediate natural surroundings and the wide variety of the planet's ecosystems that could be accessed online. This mixed reality altered their sense of self and sense of place in relation to blue spaces by generating a mind–body disconnect and digital solastalgia.

DOI: 10.4324/9781003365501-13

Digital nature: A new way of experiencing blue spaces

Constant connectivity and access to digital media have transformed human–nature relationships. From the end of the 1990s when the internet became accessible to the general public, digital nature – the online representations of actual natural environments – became a new aspect of the human relationship to blue spaces. Ochiai defines digital nature as a new perspective on nature composed of digital media:

> In digital nature, our current norms of physical and recognition abilities are transcended. Extremely enhanced computation and resolution abilities become part of daily life. The humanity of the future may live in digital nature, where the very concepts of nature, artificial objects, gravity and time are overturned.
>
> *(2019, para. 1)*

Technology is increasing our awareness of and relationship with blue environments. From waterproof smartphones and GoPros, marine navigation GPS and underwater camera traps, to drones and livestream deep-sea videos, combined with increasingly large mobile coverage, technology has helped to expand our experience of blue spaces at individual and societal levels. For many participants in my study, nature exposure involved using smart technology and staying connected in nature. Participants used technologies including mobile apps to engage with nature, for safe exploration of the environment, or for furthering their knowledge of local flora, among other things.

While Levin and Mamlok (2021) argue that technology contradicts the traditional concept of nature, Hayles (1995) asserts that the distinctions between the natural world and the artificial world are changing, and so is our perception of nature. Digital technology has become a tool to nurture our connection to blue spaces, as the following words from a participant illustrate:

> Social media allows me to learn about activities (beach clean ups, ocean swims). Documentaries inform me about the world and encourage me to engage with and protect the ocean. Apps like *Geocaching* and *iNaturalist* help me to discover and appreciate nature. My phone allows me to photograph and share the beauty of the beach.
>
> *(Study participant)*

Study participants identify multiple uses for technology in engaging with blue spaces such as education, activist, and ethical imperatives, and aesthetic appreciation. However, digital nature (including digital blue space) is a peculiar experience where the mind is online while the body is somewhere else. Digital blue spaces refer to visual, auditive, and textual components related to visible surface water experienced via digital technology. They can range from digital photographs of water bodies, short-form videos (TikTok, YouTube), and ocean documentaries, to water sounds apps, social media posts on marine ecology (Twitter, Instagram), and wildlife recordings. They offer an indirect contact with water bodies and hold the particularity of being both real and cyber, a dual reality which impacts human perception. In this dual reality, cyberspace exists jointly with physical space. The area between the two extremes, where both the real and the virtual are mixed, is called mixed reality (Milgram & Kishino, 1994), and it is this mixed reality that we experience when we go online.

Australian experiences of mixed reality and the blue

As digital users, Australian participants in my study used the internet every day, with more than half of them (55.85 per cent) connected between two and eight hours per day. Most refer to blue spaces as being part of their physical environments, citing experiences of beaches, rivers, reefs, mangroves, lakes, estuaries, or rainforest creeks. 'I grew up on the Atherton Tablelands [North Queensland], exploring the rainforest and swimming in many freshwater creeks, lakes and waterfalls with my family', a participant shared. 'I spend a lot of time by the ocean. I feel like I am in nature when swimming at the beach even if there is a city behind', another explained. However, when online, they could experience a wider variety of natural blue environments.

Digital nature, as an audiovisual rendition of nature, is not as immersive as three-dimensional virtual nature, yet it may feel more real to us because the digital natural environments either currently have a material existence with geographic coordinates, or have had in the past, in the case of natural environments that have disappeared. The physical reality of these digital natural environments cannot be ignored. Yet, the human brain treats digital information in a particular way. According to Wang and Haapio-Kirk (2021), digital media – notably through visual-centric social media platforms – is emotion-centred. For instance, Instagram pictures of ocean plastic pollution may have a visual power that can trigger emotions such as sadness, anger, or fear. Contrary to traditional media (such as newspapers, TV, and books), social media technologies are interactive. The emotional effects can be magnified when users collectively share and comment online. As Steinert and Dennis (2022, para. 2) explain, 'Those who can vividly express their emotional states in a tweet or a post have the potential to act as lightning rods that channel the collective mood'. All this shapes an understanding of digital blue spaces experienced through emotional, fragmented, and non-geographically bounded interactions.

Digital nature also enables us to redefine blue spaces in an increasingly urbanised and digitalised world and see beyond the cultural human–nature separation. As the World Health Organization states, 'Most of the earth's surface is covered by water, and most of the human body is composed of water – two facts illustrating the critical linkages between water, health and ecosystems' (2017, as cited in Britton et al., 2020, p. 50). Water is omnipresent in all expressions of nature.

Existing on earth as a solid, liquid, and gas, it can be found in water bodies, in lands, and in the air. Jue argues in *Wild Blue Media* (2020) that the human connection to nature is based on terrestrial habits of perception. These terrestrial biases prevent us from fully engaging with water bodies in general and the ocean in particular. Rethinking nature from an aquatic perspective is important because water is the common thread between all forms of nature. Human and nature have often been perceived as separate through cultural concepts such as human–nature dualism (Morton, 2010), but because blue spaces are ubiquitous and can be found everywhere, from Australian cities to the wilderness, they undermine this human–nature dualism. Nature thus bypasses the Western view of a human–nature and city–nature separation. Hall et al. (2020) confirm that it is essential to define blue spaces as including natural *and* urban settings (such as harbour front parks, rivers, ponds, lakes, ports, canals, and fountains). To this, we can now add digital blue spaces. Digital nature offers new experiences of blue spaces. However, because the body is not physically experiencing the perceived digital natural environment, this destabilises our sense of self in the context of cyberspace.

Sense of self in cyberspace: Digital blue spaces and the mind–body disconnect

> There is nothing like actually being outside with other living beings, not just pictures of them. Digital-based natural experience can be inspirational at times, but actually connecting with nature physically (outdoors) is the most rewarding. I think the digital experience eventually leads to disconnection from nature and the body.
>
> *(Study participant)*

These words from a participant highlight both positive and negative aspects of digital nature. Digital technology offers innumerable nature experiences, but the online world may separate the user from the physical experience of their immediate surroundings. One of the main findings of my study is that digital nature, while it can be intellectually and emotionally engaging, is also physically disconnecting. It results in a mind–body disconnect, which affects our sense of self, and in turn our connection to blue and non-blue spaces. Of the ten Australian participants who took part in the interviews, all agreed that screen time could cause a disequilibrium between mind and body and impact sense of self. 'When I experience nature online, I feel like I can't "access" my physical body. I don't feel connected to my physicality the way I do when hiking, swimming in the sea or exploring', one participant explained. Sense of self can be described as the subjective experience of being an *I* or a *self* bound to a body distinct from the world and others. So, why does digital technology affect our sense of self? Because the self is first and foremost situated within a body. It is born from an embodied self-awareness as we use our physical sensations to navigate the world and it involves a representation of an owned body. This is what Gallagher calls our 'body image' (2001, p. 147) or Tsakiris our 'bodily self' (2017, p. 597). This bodily self or image is challenged by the fact that we live in a mixed reality, interacting with the physical and the digital realms simultaneously. As Tsakiris explains, 'processes that update or disrupt the awareness of our physical self may in turn disrupt the physiological regulation of the self' (2017, p. 602). Studies show that digital and virtual technology can affect the perceptual processes that produce the feeling of ownership of one's body (Lenggenhager et al., 2007).[2] Van den Bos and Jeannerod (2002) also argue that movement is an important part of our sense of self. Recognising oneself requires awareness of one's body and one's actions. We identify the body as ours because we are able to move our body and observe its response to our mental commands. Thus, we may hypothesise that online time, which has attention diversion qualities and is mostly a sedentary activity, will affect our connection to our body. Additionally, living in a mixed reality can lead to distorted time perception (Gonidis & Sharma, 2017) and a different experience of space and presence (Lombard & Ditton, 1997). As this participant observes:

> Experiencing nature through a screen does not make me feel physically good, but it is a positive mental experience for the time you are watching it. I tend to forget myself, to lose track of time. Often, I stay up too late and don't get enough rest, engaging my mind but ignoring my body.
>
> *(Study participant)*

Ciaunica et al. (2022) evoke 'digital depersonalisation' to account for the dissociative mind–body experience induced by online time. Investigating the relationship between digital media-based activities, distal social interactions, and people's sense of self during lockdown, they found that increased use of digital technology correlated with higher feelings of depersonalisation and negative emotions. Depersonalisation is a common dissociative experience characterised by distressing feelings of being detached from one's self, body, and/or the world. While what participants in my study experienced cannot be described as a psychological disorder, their reports of 'feeling unreal', 'losing myself', or 'forgetting everyone around' involve a mild form of digital depersonalisation, which, in turn, can affect relationships with blue spaces.

In light of this, can digital blue spaces provide a whole human–nature experience since one is just mentally – and not physically – engaged with the blue? And can online blue activism be effective in this respect? The majority of my study's participants used digital technology as a tool for eco-activism and to learn about conservation debates related to blue environments. For this participant, social media was seen as a positive tool to protect the ocean:

> I'm on social media, and I read green left weekly newspapers or journals of social actions. Just like nature, social media broadens my landscape of interactions

> and engagement, and inspires me with doable solvable actions to help save the ocean. I love petitions and writing letters to politicians.
>
> *(Study participant)*

Emerging theories on ocean identity and marine citizenship emphasise the importance of self-identity to effectively care for blue environments. They also show that a physical and sensory engagement with marine environments is necessary. The concept of ocean identity (Jaksha, 2019) – which is located within the broader concept of environmental identity (Clayton, 2003) – posits that the cognitive, emotional, and behavioural dimensions of our identity are essential to support healthy human–ocean relationships and eventually pro-environmental behaviours. The cognitive dimension of ocean identity refers to how one intellectualises their relationship with the ocean, the emotional dimension can be understood as one's sense of care for the ocean, and the behavioural dimension involves one's actions to protect the ocean (for instance, avoidance of single-use plastics or participation in a beach clean-up). Similarly, the concept of marine citizenship (McKinley & Fletcher, 2012) describes the rights and responsibilities of an individual towards the marine environment. Buchan et al. (2023) explain that a marine conscience and a marine identity are important in fostering a strong sense of citizenship responsibility and moral duty. As Buchan explains:

> For people who depend upon a healthy ocean for their identity, threats to the ocean, such as climate change, pollution and loss of habitats and biodiversity, also threaten their personal identity. People are motivated to protect their identity so this could drive them to take action, becoming active marine citizens and trying to reduce those ocean threats.
>
> *(2021, para. 9)*

But when the blue environment is experienced digitally, it lacks the physical engagement that actual nature provides. Buchan et al. (2023) argue that a physical connection with place is key in incorporating a knowledge and an understanding of it. It drives the motivation to take action to protect that place. Repeated physical interaction with the ocean is understood to lead to interest, education, and the development of an ocean identity. Jue confirms, through her personal recollections of diving, that physical experiences of aquatic environments may give rise to a 'planetary self' (2020, p. 74) who is more aware of the affective dimension of the human–nature bond and more ethically engaged with the environment. She adds that, from the position of being submerged in water, one can fully comprehend the differences between oceanic and terrestrial experiences of space, time, and embodiment. This may point to the limitations of digital nature to render the richness and complexity of the blue space experience. From an intellectual and emotional standpoint, digital technology may increase our awareness of blue environments, but it cannot provide a bodily knowledge, thus reinforcing a mind–body disconnect. This leads us to question sense of place in the context of cyberspace, and to acknowledge the ecological distress that it can engender and which I call digital solastalgia.

Digital solastalgia and online place attachment

Digital solastalgia – a term that I coined based on Glenn Albrecht et al.'s research (2007) – refers to environmentally induced distress experienced through digital technology and digital media. The place(s) which we call *home* have been redefined by the introduction of the internet in our private dwellings, expanding our experience of nature and our sense of place to the rest of the world. Albrecht (2019) developed the concept of solastalgia to express a form of nostalgia for a natural home environment that has been negatively changed. He offers the following definition:

> Solastalgia is a new concept developed to give greater meaning and clarity to environmentally induced distress. As opposed to nostalgia – the melancholia or homesickness experienced by individuals when separated from a loved home – solastalgia is the distress that is produced by environmental change impacting on people while they are directly connected to their home environment.
>
> *(Albrecht et al., 2007, p. 95)*

Like solastalgia, which Albrecht refers to as the homesickness that you have at home, digital solastalgia pertains to our psycho-emotional bond to nature in relation to our home environment and the environmental destruction that we witness in reality and via digital technologies. Other experiences of ecological distress such as eco-anxiety and eco-grief are also burgeoning (Hayes & Poland, 2018). Eco-anxiety relates to the pain resulting from living in anticipation of climate change-driven environmental transformation and witnessing environmental destruction elsewhere on the planet (Popescu, 2022). Similarly, eco-grief is defined by Cunsolo and Ellis (2018) as a grief associated with physical ecological losses. My research indicates that, via digital technology, our home environment can be extended to remote natural environments. As a result, we can feel solastalgia for places that we have never engaged with outside of online spaces, but which we feel are virtually connected to our own immediate home environments. Digital nature can generate a shift from a local home environment to a global one, and a sense that, as a 'planetary self', the planet is our home. In this respect, digital solastalgia is different from eco-anxiety because it is related to the notion of home, and it is different from eco-grief because there is no physical experience of loss, only a digital one.

Digital solastalgia may be a growing problem for our connected society since constant connectivity makes it difficult to avoid hearing about environmental issues. As a participant points out, the environmental crisis 'can be beamed into your world every day, multiple times a day, along with the human distress it causes'.

But digital media also deepens our understanding of two specific issues in relation to blue spaces: climate change and plastic pollution. The following words from a participant recounting her experience of solastalgia exemplify this:

> I mostly feel [solastalgia] on beaches that are littered with plastics, and when people use unnecessary single use plastics like disposable coffee cups, straws, and so on. I also feel it through social media and digital news (i.e. climate change), species under threat due to habitat destruction and the most recent one is the oil spill in the Solomon Islands. It causes me to feel a great amount of grief, which is why I do everything I can to minimise my plastic and carbon footprints.
>
> *(Study participant)*

Digital media is helping to increase climate change awareness and, as a result, boost the public perception of blue environments (Baiardi & Morana, 2020). With millions of tonnes of plastic entering the oceans every year (IUCN, 2021), information campaigns – such as *Plastic Free July* or *The Last Straw* in Australia – have encouraged people to reduce plastic consumption. Social media, as one of the most prevalent forms of communication, plays a role in these campaigns. According to Rapada et al. (2021), social media can influence the consumption behaviour of individuals towards plastic pollution and can be effective in supporting marine protection. Taking action to reduce plastic consumption can alleviate some of the distress induced by digital solastalgia, which, as many study participants explained, is generally linked to a feeling of being powerless, and an impression that nothing can be done.

However, digital media, while supporting awareness of blue environments and providing a means to advocate for marine protection, also fosters ecological distress and questions our sense of place in relation to cyberspace. Sense of place is a subjective mental and emotional perception of the environment and the experience of being in a particular point in space (Erfani, 2022). In a digital context, it is directly linked to the *presence* of the user:

> Presence (a shortened version of the term 'telepresence') is a subjective perception in which even though part or all of an individual's current experience is generated by and/or filtered through human-made technology, part or all of the individual's perception fails to accurately acknowledge the role of the technology in the experience.
>
> *(ISPR, 2000, para. 1)*

The immersiveness of digital technology initiates what Lombard and Ditton describe as 'the perceptual illusion of nonmediation' (1997, para. 1). This sense of presence creates an illusion of place. The degree to which digital technology can produce seemingly accurate representations of blue spaces generates a feeling of transportation, of being a part of the blue environment. Sense of place is born when the digital user attaches meaning and affection to the natural environment, regardless of the environment being mediated by technology or not. This shows how one's attachment for a place can then be extended to digital blue environments. And it may account for digital solastalgia since solastalgia is positively correlated with place attachment. Phillips and Murphy (2021) confirm that those having higher levels of place attachment express stronger feelings of solastalgia.

Cyberspace offers a digital representation of nature within a computer space (Thomas, 2013). In the process of becoming attached to remote natural environments experienced online (and failing to see the technological mediation), we may become attached to the very thing that makes that connection possible: the digital device. Online place attachment can make us view a smartphone as a portable sense of place. A safe place we go back to when we do not want to deal with what is right in front us, a way to both connect to the world and escape our immediate surroundings, and a way to stay in touch with blue spaces. For instance, a participant in my study would listen to ocean sounds via a free app to relax whenever she had to take public transport to town. The digital age, which, as Wiegel (2010) says, marks the dominance of the disembodied eye (because we are viewing the world through our screens rather than physically experiencing it), has been described as a step away from our body and from a sensory experience of nature. But could the disembodied experience resulting from the internet – or the location-independent lifestyle embraced by digital nomads and made possible by new technologies which redefine place attachment – paradoxically be a way to return to the body? Ultimately, our sense of place starts within ourselves. Our first home is our body. In a world that is constantly changing, moving, and expanding, where natural environments are being threatened on local and global scales, going back to the body (even through digital spaces and devices) can feel like going home. Digital technology is a powerful tool and it must be used in a balanced and intentional manner to moderate the mind–body disconnect. Digital blue spaces should remain an inspiration to go outside and engage with real blue spaces in order to reconnect to our body.

Digital technology challenges us to experience blue spaces in new ways. It can provide online access to remote blue spaces and increase our awareness of aquatic issues such as climate change and plastic pollution. But it can also impact how we relate to our own body, generate a mind–body dissociative experience, and engender digital solastalgia. Digital technology makes it possible for our sense of self to extend to remote blue environments and fosters an expansion of our environmental identity to embrace virtually experienced spaces as an extension of our home environment. But digital nature, being at best an audiovisual representation of nature, restricts our physical and sensory experience of the blue. Digital blue spaces do not offer a direct physical experience, instead they raise questions about using technology to sustain marine identity and pro-environmental behaviours. Thus, while digital technology can be an alternative to nurture relationships with blue spaces, if we do not want to diminish our experience of the blue, we have to keep in mind that it has limitations and employ it not as a substitute, but as a way to enrich actual engagements with blue space.

Notes

1 The project was approved by the James Cook University Human Research Ethics Committee (reference number: H7105) and complies with the ethical standards of the institution.
2 Lenggenhager et al. (2007) induced a full-body illusion by having participants view a virtual body presented at a distance of 2 metres ahead of them through the use of a 3D-video head-mounted display. After synchronous stimulation, participants felt as if the virtual body was their body.

References

Albrecht, G. A. (2019). *Earth Emotions: New Words for a New World.* Cornell University Press.
Albrecht, G., Sartore, G. M., Connor, L., Higginbotham, N., Freeman, S., Kelly, B., Stain, H., Tonna, A., & Pollard, G. (2007). Solastalgia: The distress caused by environmental change. *Australasian Psychiatry, 15*(1), 95–98.
Baiardi, D., & Morana, C. (2020). Climate change awareness: Empirical evidence for the European Union. *University of Milan Bicocca Department of Economics, Management, and Statistics Working Paper, 426*, 2–33.
Britton, E., Kindermann, G., Domegan, C., & Carlin, C. (2020). Blue care: A systematic review of blue space interventions for health and wellbeing. *Health Promotion International, 35*(1), 50–69.
Buchan, P. (2021, 8 July). Marine 'identity' could boost ocean protection. University of Exeter – Green Futures. https://greenfutures.exeter.ac.uk/article/marine-identity-could-boost-ocean-protection/
Buchan, P. M., Evans, L. S., Pieraccini, M., & Barr, S. (2023). Marine citizenship: The right to participate in the transformation of the human–ocean relationship for sustainability. *PLOS One, 18*(3), Article e0280518.
Ciaunica, A., McEllin, L., Kiverstein, J., Gallese, V., Hohwy, J., & Wozniak, M. (2022). Zoomed out: Digital media use and depersonalization experiences during the COVID-19 lockdown. *Scientific Reports, 12*, Article 3888.
Clayton, S. (2003). Environmental identity: A conceptual and an operational definition. In S. Clayton & S. Opotow (Eds.), *Identity and the Natural Environment: The Psychological Significance of Nature* (pp. 45–65). MIT Press.
Cunsolo, A., & Ellis, N. R. (2018). Ecological grief as a mental health response to climate change-related loss. *Nature Climate Change, 8*(4), 275–281.
Erfani, G. (2022). Reconceptualising sense of place: Towards a conceptual framework for investigating individual–community–place interrelationships. *Journal of Planning Literature, 37*(3), 452–466. https://doi.org/10.1177/08854122221081109
Gallagher, S. (2001). Dimensions of embodiment: Body image and body schema in medical contexts. In K. Toombs (Ed.), *Handbook of Phenomenology and Medicine* (pp. 147–175). Kluwer Academic Publishers.
Gonidis, L., & Sharma, D. (2017). Internet and Facebook related images affect the perception of time. *Journal of Applied Social Psychology, 47*(4), 224–231.
Hall, K., Garrett, J. K., White, M. P., Grellier, J., Wuijts, S., & Fleming, L. E. (2020). Using urban blue spaces to benefit population health and wellbeing. The Blue Health Project. https://bluehealth2020.eu/wp/wp-content/uploads/2020/11/Read-about-the-benefits-of-blue-spaces_BlueHealth-Project_Horizon-2020.pdf

Hayes, K., & Poland, B. (2018). Addressing mental health in a changing climate: Incorporating mental health indicators into climate change and health vulnerability and adaptation assessments. *International Journal of Environmental Research and Public Health, 15*(9), 1806–1825.

Hayles, K. (1995). Simulated nature and natural simulations: Rethinking the relation between the beholder and the world. In W. Cronon (Ed.), *Uncommon Ground: Rethinking the Human in Nature* (pp. 409–425). W. W. Norton.

ISPR. (2000). The concept of presence: Explication statement. International Society for Presence Research. https://ispr.info/about-presence-2/about-presence/

IUCN. (2021, November). Issues brief: Marine plastic pollution. IUCN. https://www.iucn.org/sites/default/files/2022-04/marine_plastic_pollution_issues_brief_nov21.pdf

Jaksha, A. P. (2019). Leveraging ocean identity in education to impact students' conservation practices. In G. Fauville, D. Payne, M. Marrero, A. Lantz-Andersson, & F. Crouch (Eds.), *Exemplary Practices in Marine Science Education* (pp. 93–107). Springer.

Jue, M. (2020). *Wild Blue Media: Thinking through Seawater.* Duke University Press.

Lenggenhager, B., Tadi, T., Metzinger, T., & Blanke, O. (2007). Video ergo sum: Manipulating bodily self-consciousness. Science, *317*(5841), 1096–1099.

Levin, L., & Mamlok, D. (2021). Culture and society in the digital age. *Information, 12*, 68–81.

Lombard, M., & Ditton, T. (1997). At the heart of it all: The concept of presence. *Journal of Computer-Mediated Communication, 3*(2). https://doi.org/10.1111/j.1083-6101.1997.tb00072.x

McKinley, E., & Fletcher, S. (2012). Improving marine environmental health through marine citizenship: A call for debate. *Marine Policy, 36*(3), 839–843.

Milgram, P., & Kishino, F. (1994). A taxonomy of mixed reality visual displays. *IEICE Transactions on Information and Systems, 12*(12), 1321–1329.

Morton, T. (2010). *The Ecological Thought.* Harvard University Press.

Ochiai, Y. (2019). The form of digital nature. *Ars Electronica.* https://ars.electronica.art/outofthebox/en/digitalnature/

Phillips, C., & Murphy, C. (2021). Solastalgia, place attachment and disruption: Insights from a coastal community on the front line. *Regional Environmental Change, 21*, 46–60. https://doi.org/10.1007/s10113-021-01778-y

Popescu, C. (2022). Eco-anxiety and solastalgia: Keeping hope alive. *Rhizome, 82*(1), 11–12. https://doi.org/10.3917/rhiz.082.0011

Rapada, M. Z., Yu, D. E., & Yu, K. D. (2021). Do social media posts influence consumption behavior towards plastic pollution? *Sustainability, 13*(22), Article 12334. https://doi.org/10.3390/su132212334

Steinert, S., & Dennis, M. J. (2022). Emotions and digital well-being: On social media's emotional affordances. *Philosophy and Technology, 35*, 36–57.

Thomas, S. (2013). *Technobiophilia: Nature and Cyberspace.* Bloomsbury.

Tsakiris, M. (2017). The multisensory basis of the self: From body to identity to others. *Quarterly Journal of Experimental Psychology, 70*(4), 597–609. https://doi.org/10.1080/17470218.2016.1181768

Van den Bos, E., & Jeannerod, M. (2002). Sense of body and sense of action both contribute to self-recognition. *Cognition, 85*(2), 177–187. https://doi.org/10.1016/S0010-0277(02)00100-2

Wang, X., & Haapio-Kirk, L. (2021). Emotion work via digital visual communication: A comparative study between China and Japan. *Global Media and China, 6*(3), 325–344.

Wiegel, L. (2010). *Perception in the digital age: Analysing aesthetic awareness of changing modes of perception* [RMA Thesis, Utrecht University]. https://studenttheses.uu.nl/bitstream/handle/20.500.12932/5550/thesislw.pdf?sequence=2&isAllowed=y

10

'A DANCING CREATURE OF CRIMSON AND YELLOW'

Writing the Great Barrier Reef

Jessica White

My first encounter with the Great Barrier Reef was in a glass-bottomed boat near Brampton Island in the mid-1980s. For a small girl from the country who only saw the sea once a year in the summer holidays, and who was used to the contained and chlorinated water of swimming pools, the purple and red coral, shoals of sparkling fish, and the ruffled mouths of giant clams were astonishing. Just as it was difficult for an 8 year old to comprehend the foreignness of the Reef's creatures, so too is it hard to take in the magnitude of the Reef itself. The Great Barrier Reef is, as Iain McCalman writes in *The Reef: A Passionate History*, 'so extensive that no human mind can take it in, the exception perhaps being astronauts who've seen its full length from outer space' (2013, p. 3). The world's largest living structure, it consists of 2,500 reefs and 900 islands, and is home to nearly 9,000 species of marine life (Bellwood and Mallela, 2003). Its beauty and complexity have drawn explorers, scientists, writers, tourists, and photographers to its atolls and cays for centuries.

In time, I would grow up to become an author, and my early memories of swimming in a foreign, salty world filtered into my writing. In 2022, the Foundation for Australian Literary Studies partnered with the Australian Institute of Marine Science (AIMS) to host six writers at AIMS's base near Bald Reef, Townsville (James Cook University, 2022, 'Event'). As one of the writers, I learned about AIMS's research to breed corals that can withstand warming temperatures. These experiments were conducted through the National Sea Simulator, or SeaSim, which, through a series of seawater tanks, enables scientists to conduct large, long-term experiments on the impact of environmental change on corals and other marine organisms. I incorporated this information into my work-in-progress, a speculative fiction novel about sea creatures and the increasing ocean temperatures of the Great Barrier Reef.

DOI: 10.4324/9781003365501-14

My work is part of a lineage of Australian literature that has responded to the Reef in a range of genres. As Iain McCalman notes, the Reef was 'built by human minds as well as by coral polyps' (2013, p. 9). A search for the phrase 'Great Barrier Reef' in AustLit, the Australian literature database, reveals 910 entries from 1873 to 2022.[1] However, scholarly critique of this writing, and what it might signify, remains scanty. Pierce's survey of the field focuses on North Queensland rather than the Reef specifically, and the most recent text he cites is Maxwell Grant's *Barrier Reef* (1980). Stephen Torre, long-time editor of *eTropic: Electronic Journal of Studies in the Tropics*, has published essays (2005 and 2008) on Di Morrissey's *The Reef* (2008) as well as the journals of James Cook, Mark O'Connor's *Reef Poems* (1976), Rosaleen Love's *Reefscape: Reflections on the Great Barrier Reef* (2000), and Naomi Mairou's children's book *The Dugong Meadow* (2002). Numerous critics have focused on the writing of E. J. Banfield, long-term resident of Dunk Island (Ross, 1997; Perchan, 2000; Zeller, 2002, 2004, and 2007; Taylor, 2004; Weaver, 2015), while some have attended to Jean Devanny's (Taylor, 2004) and Dorothy Cottrell's (Ross, 1997) writing on the Reef. To date, poet Judith's Wright's iconic work *The Coral Battleground* has received no attention from literary studies scholars beyond reviews, which seems remiss given that she was one of the trio whose activism and intellectual energy saved the Reef from mining.

This chapter explores representations of the Reef in Aboriginal and Torres Strait Islander peoples' writing, nature writing, political writing, and children's literature by Australian writers who have lived on or near the Reef. It aims to chart how writing about the Reef has changed over time, particularly as advances in science have illuminated threats to the Reef from mining, agriculture, climate change, and the Crown of Thorns starfish. It also reflects on a shift in focus from human protagonists to nonhuman protagonists. This shift is, arguably, a reflection of posthumanism, which questions 'what exactly is the basic unit of common reference for our species, our polity, and our relationship to other inhabitants of this planet' (Braidotti, 2013, pp. 1–2). Astrid Neimanis, in *Bodies of Water* (2017), suggests that water is one of these units. She observes that 'the flow and flush of waters sustain our own bodies, but also connect them to other bodies, to other worlds beyond our human selves' (Neimanis, 2017, p. 2). Recent writing, in particular, underscores the critical importance of the Reef for humans and nonhumans alike by dwelling on humans' embeddedness in and dependency on the Reef.

Sea Country

The first humans to watch the Reef form were Aboriginal and Torres Strait Islander peoples. The Reef's present structure is estimated to be 7,000 years old, and before the seas rose and the Reef formed, First Nations peoples lived on what is now the sea floor. They have cared for Sea Country since this time, and there are more than seventy Aboriginal and Torres Strait Islander Traditional Owner groups that have continuing relationships with the Reef region and its natural resources ('Reef Traditional Owners'). The term 'Sea Country' is used by Aboriginal and Torres Strait Islander peoples to describe both land and waters over which they have custodianship. They consider the sea 'as much a part of the country as land estates' and it is a 'key component of their identity' (Perrett, 2010, p. 6). As the Woppaburra peoples of the Keppel Islands explain, 'we have a lifelong spiritual and physical connection to the land and the sea . . . (and) a lifelong responsibility to care for land and sea country' (cited in Perrett, 2010, p. 6).

Aboriginal and Torres Strait Islander peoples' stories are part of an oral tradition reaching back tens of thousands of years. Traditional custodians from the North Wellesley Islands, the Narrangga people of the Yorke Peninsula, and the Yidindji people of the Cairns area describe islands that were once connected to the mainland (Reid & Nunn, 2015). Gurang and Gooreng Gooreng tour leader Cherissa Blackman describes a Gimoy Yindinji story of how two brothers were told not to hunt a particular fish by Bhiral, the Creator. One of the brothers ignored this directive and speared the fish. Bhiral, who was angry, threw lava and hot rocks from the sky, which caused the sea levels to rise (Stünzner, 2018).

While these oral stories are not captured by the AustLit database, which focuses on published works, Aboriginal and Torres Strait Islander peoples have detailed their relationships with the Reef and its inhabitants in print form. Artist Munganbana Norman Miller's *Reef and Rainforest: An Aboriginal Voice through Art and Story* (2015) collects over two hundred of his artworks. Munganbana, who is a member of the Jirrbal, Bar-Barrum, and Tableland Yidinji groups and whose name means 'mountain water' in the Jirrbal language (Miller, 2015, p. vi), has written a few lines of explanatory text to accompany each image. Often these are simple and informative, but occasionally they draw on the poetic. 'Dancing Stream Turtles Diptych' features five streams of water converging at a circle. Seven turtles are interspersed among the streams, their backs patterned like shields. The colours of the painting are ochre, black, white, and yellow, with touches of pale blue. Of this work Munganbana writes:

> I could see the turtles dancing, enjoying life celebrating, and I envisaged them being drawn towards streams of flowing water, almost into a vortex in the centre where there is a whirlpool and they blend into one. The turtles' fan-shaped headdresses and stylized legs are painted up in earthy colours, so they are ready to water-flow corroboree.
>
> *(Miller, 2015, p. 78).*

The illustration and accompanying story demonstrate a mingling of human and nonhuman, with the turtles taking on the human attributes of corroboree and ceremony, and the land and sea merging through the use of colour, with ochre denoting land and blue denoting the sea. The work echoes Neimanis's observation, cited in the Introduction, about the ways that water connects human and non-human. It also highlights Munganbana's awareness of his inseparability from his Country, as he writes, 'I paint from who I am and from the rich natural and cultural landscape that I am a part of' (Miller, 2015, p. vi). Just as the turtles and water 'blend into one', so too are humans part of a continuum that connects them with the Reef and its inhabitants.

'We get plenty of writers'

Stephen Torre's 2008 analysis of literature about the Reef opens with shipwrecks, delving into Captain James Cook's and Joseph Banks' journal entries from 1770. Torre observes that the journals 'provided ground-breaking accounts of the geographical details, Indigenous people, and flora and fauna of the continent' but 'largely ignored the Reef except for reinforcing its perils for navigators'. Much the same might be said of Frank Reid's *The Romance of the Great Barrier Reef* (1954), a collection of vignettes about encounters on the Reef, with many mentions of 'savages', black and white alike, murder, and wrecks.

Cheryl Taylor notes that some writers of the earliest twentieth century, rather than perceiving the Reef as a barrier, instead saw it as a place of liminality, a hub 'where the romantic trope of tropical island escape . . . intersects with nature writing and with travel and tourist literature' (2004, p. 18). The change in focus may have come about because settler writers such as Edmund James Banfield and Jean Devanny came to the Reef for recuperation, rather than traversing it by ship, and stayed for extended periods.

Dunk Island, known as 'Coonanglebah' in the Warrgamay and Dyirbal languages, is famed for its association with the writer E. J. Banfield. Hailing from Victoria and in the grips of a nervous breakdown as deputy editor of the *Townsville Bulletin* newspaper, Banfield arrived at Dunk Island in September 1897 with his deaf wife Bertha. Dunk Island sits off the coast of Mission Beach and Bingil Bay in Far North Queensland, which Judith Wright would later write of in her book *The Coral Battleground* (2014). Even today, Dunk Island can easily be viewed from the home of John Büsst, an artisan activist who led the first Save the Barrier Reef campaign, near Mission Beach, Far North Queensland.

In *The Confessions of a Beachcomber*, Banfield records arriving at the island as a 'sick man caring for naught in his physical collapse and disconsolation' (1908, p. 11). On waking the next morning, however, he is immediately restored by

> a perfect combination of invigorating elements. The cloudless sky, the clear air, the shining sea . . . the sweet odours from the eucalypts and the dew-laden grass, the luminous purple of the islands to the south-east; the range of mountains to the west and north-west, and our own fair tract — awaiting and inviting.
>
> *(Banfield, 1908, p. 11)*

As Pierce (2004) has noted, the Djiru people had fished on the island for thousands of years, but Banfield claimed it for himself through lease and, later, selection. His love for and detailed attention to the natural world yielded luscious, sensory-laden sections of prose. While the above observations draw upon the olfactory, his descriptions of the Reef are, necessarily, largely visual. For example, when describing 'gardens of coral', his prose is crammed with adjectives:

> Rounded, skull-like protuberances with the convolutions of the brain exposed, stag-horns, whip-thongs yards long, masses of pink and white resembling fanciful confectionery, intricate lace-work in the deepest indigo blue, have their appointed places. Some of the spreading plant-like growths are snow-white, tipped with mauve, lemon-coloured tipped with white, white tipped with lemon and pale blue.
>
> *(Banfield, 1908, p. 130)*

Banfield's long sentences of double-barrelled words, saturated with colour and references to sweetness and delicacy, are a luscious paean to the abundance of marine life. This is not to say, however, that he romanticises the Reef's inhabitants entirely. In his next breath, he describes a coral reef as 'one of Nature's most cruel battlefields' in which 'molluscs are murderers and the most shameless of cannibals. No creature at all conspicuous is safe, unless it is agile and alert, or of horrific aspect, or endowed with giant's strength, or is encased in armour' (Banfield, 1908, p. 130). While the sheer weight of adjectives in his prose makes for slow reading, it also gestures to the abundance of species in the Reef's ecosystems.

Some forty-five years after Banfield's presence on the Reef, Jean Devanny, a writer born in New Zealand who became a committed member of the Australian Communist Party, travelled up the coast of Queensland. A champion of workers' and women's rights, she met and wrote about those who made their living from the sea and forests (see also Killian Quigley's exploration of Devanny and the Reef in Chapter 6 of this collection). For Devanny, as for Banfield, the journeys to Queensland were ones of restoration, following personal losses and the demands of political life (Taylor, 2004, p. 19).

As well as writing of the humans she encountered, Devanny described some of the nonhumans in the area such as 'forget-me-not blue' female soldier crabs (1944, p. 154). Her description of the cays in *Travels in North Queensland* (1951) is 'the longest account of the reefscape in Devanny's published writings' (Taylor, 2004, p. 19). In *By Tropic Sea and Jungle: Adventures in North Queensland*, Devanny recounts a conversation with an elderly oyster farmer living on a shanty on Dunk Island, in which he described Banfield as

> "A queer old stick. Nice to you for a few minutes, then let you know he wanted you to go. His wife liked to be alone, too. They were well matched. A real distaste for strangers . . . He saw things real crooked".
>
> *(1944, p. 31).*

The elderly farmer also mused, '"I reckon Frank Clune is good. He was up here, you know. Vance Palmer was here, too . . . We get plenty of writers"' (Devanny, 1944, p. 32).

Another of these writers was Dorothy Cottrell, who read Banfield's *Confessions of a Beachcomber* as a young woman and was so enamoured that her mother had 'once tried to arrange for Dorothy to visit the reef and to holiday on Banfield's island' (Ross, 1997, p. 57). After secretly marrying the bookkeeper for her family's property at Ularunda, Walter Cottrell, and carrying out a clandestine correspondence with Banfield in 1922, the Cottrells arranged to live in Banfield's spare cottage on the island. In February the next year, the couple moved to Dunk Island without telling Dorothy's parents. Dorothy, who used a wheelchair after contracting polio as a child, describes her experiences of the Reef in her first novel, *The Singing Gold*:

> There was life and colour and light, and life, life, life in the glittering tropical air quivering with the diamond fire of noon, in the vine-rioting jungle on the orchid-knotted rocks; life rising from the coloured floor of the sea in pillars and reefs and flower-forms of coral.
>
> *(1928, p. 81)*

As with Banfield, in his observation that 'in these fervid and fecund waters life is real, life is earnest' (1908, p. 130), Cottrell uses repetition to underscore the multiplicity of life in the Reef, while her deployment of commas and semi-colons gestures to the way that living creatures – and the Reef itself – are interconnected.

The Reef drew writers because of its obvious beauty, sensory satiation, and escape from the pressures of family and work. These three writers dwelled on its aesthetic qualities, to which their imaginations responded with vibrancy and poetry. However, when it came to a fight for the Reef's life some three decades later, one of Australia's most famous poets chose not to write about aesthetics, but on politics, legislation, and science.

Literary activism

Judith Wright's *The Coral Battleground*, first published in 1977, is 'the story of the battle of a few people who loved it; the battle to save that thousand-mile stretch of incomparable beauty from the real destroyers – who are ourselves' (2014, p. xxvi). Wright, one of Australia's most loved and eminent poets, steered away from her conservative pastoral heritage and became an activist for Aboriginal and Torres Strait Islander peoples' rights and an environmentalist. One of the founders of the Wildlife Preservation Society of Queensland in 1962, she was contacted by artist John Büsst in 1967 to support an objection to the public hearing on a mining application to mine Ellison Reef off the coast at Innisfail.

Büsst, driving a campaign that drew on environmentalists' personal networks and encouraged the media to focus on scientific evidence (Lloyd et al., 2017, p. 57), successfully prevented the mining application from being approved. Despite this, in 1967 Queensland's Minister for Mines and Main Roads introduced a bill that made it possible to drill the Reef for oil (Lloyd et al., 2017, p. 57). It also emerged that the Queensland premier, Joh Bjelke-Petersen, had investments in oil companies that held leases over the Reef (Lloyd et al., 2017, p. 58). Büsst was friends with the then Prime Minister Harold Holt, who 'was a convert to John's ideas on the need for protecting the Reef' (Wright, 2014, p. 4). Although Holt assured Büsst that the federal government would take over the Reef (Lloyd et al., 2017, p. 58), a drilling rig was permitted by the Queensland government to arrive in 1970. In response, a Royal Commission into Exploratory and Production Drilling for Petroleum in the Area of the Great Barrier Reef was established by the Queensland and federal governments (Lloyd et al., 2017, p. 59). At its conclusion four years later, the new Whitlam government passed an act to seize 'federal sovereignty over Australia's submerged resources' (Lloyd et al., 2017, p. 59), and a year later it introduced the Great Barrier Reef Marine Park Act 1975. In 1981, the Reef became a UNESCO World Heritage site.

Wright's account of Büsst's, Len Webb's, and her actions to save the Reef is, in her own words, an 'unadorned, bare chronological account' (2014, p. 187). Rather than dramatise encounters, she chose to 'give the facts and little more' (Wright, 2014, p. 187). She describes Commonwealth and state legislation, court appearances, scientists and research bodies with whom the activists engaged, as well as emerging scientific research. Dr Len Webb, a forester who joined the Queensland Wildlife Society and who undertook one of Australia's first doctorates in the new subject of ecology (McCalman, 2017), was critical to this research, as was Eddie Hegerl, an experienced diver and collector, who identified coral, mollusc, and fish species to prove that the Reef, as was commonly assumed, was not 'dead' (Wright, 2014, p. 11). Only at the time of Wright's campaigning did people become aware that the Reef was an interconnected structure. Wright explains that to talk of the Reef 'is to talk of many hundreds or thousands of reefs; yet it is also to speak of what is now being increasingly recognised as an ecological unity' (2014, pp. 20–21). The water, tidal movements, and winds, she surmises, 'are probably all indispensable to the maintenance and dynamics of its living species' (Wright, 2014, p. 20) and could be disturbed by blasting channels or creating harbours for boats. Research into the complex Reef ecosystem was until this point not a priority area of research in Australia, and little was known about how the marine and terrestrial zones interacted (Wright, 2014, pp. 22–23, p. 117). The Save the Reef campaign made politicians and the general population better understand how the different corals, cays, and ocean life were fundamentally interconnected. Scientists from America also provided first-hand knowledge of an oil spill in Santa Barbara, disputing the concept that drilling for oil in one particular area would not have an impact somewhere else (Wright, 2014, p. 51). The combination of new scientific evidence, a globalising environmental movement, and a very real threat of mining on the Reef provided Wright with the knowledge to draw on and knit together in *The Coral Battleground.*

However, her attention to the intricacies of the environmental campaign required a different literary style, and Wright's poetic attention to the natural world, expressed so vividly in poetry collections such as *Woman to Man* (1949), *The Two Fires* (1955), and *Birds* (1962), appears only briefly in *The Coral Battleground.* She describes her early encounters with the Reef, the 'marvellous blue calm of waters, the green island of the Whitsunday passage rising out of them' (Wright, 2014, p. 37). At her book's close, she includes one section of prose that describes a rockpool:

> sea-anemones drifted long white tentacles above the clean sand, and peacock-blue fish, only inches long, darted in and out of coral branches of all shapes and colours. One blue sea-star lay on the sand floor. The water was so clear that every detail of the pool's crannies and their inhabitants was vivid, and every movement could be seen through its translucence. In the centre of the pool, as if on a stage, swayed a dancing creature of crimson and yellow, rippling all over like a wind blown shawl.
>
> *(Wright, 2014, p. 187)*

The creature, a Spanish dancer or nudibranch, was for Wright an 'inner image of the spirit of the Reef itself', one she held in her mind when she and her companions 'thought of the shadow that menaced the Reef' (2014, p. 188). The colour, movement, and delicacy of the sea creature is a beacon against the darker 'shadow' of capitalism and pollution. Aside from these few instances, the work's tone is relatively formal and factual.

Brigid Rooney, in *Literary Activists*, notes that Wright's poetry was seen to be losing its energy during the late 1960s and 1970s on account of her activism (2009, p. 19). Poet Les Murray complained that Wright had lost her compassion and her 'Dreaming' (a reference to her pastoral heritage) (qtd. in Rooney, 2009, p. 84), while her daughter Meredith McKinney suggests that activism might have been 'another way of expressing her vision at a time when her poetic energies were beginning to dissipate' (qtd. in Rooney, 2009, p. 19). Tim Bonyhady (2005) recounts the strain Wright experienced between her art and activism (although Wright identified as a conservationist, as the term 'activist' was not used until the 1960s, Bonyhady argues that Wright's social and political engagement made her one). He notes that 'in mid-1971 Brisbane's *Sunday Mail* reported that she had not written a poem for months. A year later she finally had a new collection in manuscript but poems were still only occasionally "squeezed" out of her' (Bonyhady, 2005). He suggests that *Fourth Quarter*, published a year before *The Coral Battleground*, was one of her weakest collections.

Rooney disagrees with these assessments, writing that many of Wright's poems from this period 'are remarkable exercises in restoring the poet's sense of value and self, a self otherwise exposed with eyes painfully open to the horror of a destructive humanity' (2009, p. 20). Other critics support these observations, observing that in *Fourth Quarter* (1976) Wright achieves 'a disillusioned subjectivity that presents a bleak vision of extinction' (Wallace and O'Carroll, 2016, p. 135). In 'Easter Moon and Owl', she writes that 'The lyric note may vanish from my verse' and then concludes, 'Muse, if they mock you,/you may yet grin last./Man's signs are ominous' (Wright, 1976, p. 3). Although the poetic elements of her work may disappear, she suggests, her writing will prevail, even as humankind destroys itself.

Wright's persistence with writing despite the exhaustion of campaigning indicates its centrality to her activism. Wallace and O'Carroll observe that she 'did not *simply* abandon poetry in favour of politics, but rather . . . she spent the last three decades of her life trialling different modes of articulation and different writing genres' (2016, p. 118). *The Coral Battleground* can be seen as one such articulation in the genre of prose which, as Wright argued at the 1995 Association for the Study of Australian Literature conference, was 'as important . . . as poetry, and indeed indispensable to the writing of poetry itself' (qtd in Bonyhady, 2005). This is evident in the passage describing the rockpool, with the longer and shorter sentences creating the rhythms that mark her poetry. Reviewer Kate Livett also observes that Wright's 'facility with language is the unacknowledged underside of this text' (2015, p. 215). Through her literary skill, Wright 'creates a narrative that delivers the relevant scientific, legal, political, financial and bureaucratic detail smoothly within the broader narrative arc that pushes the reader along' (Livett, 2015, p. 216). Wright's command of structure and detail, honed from years of writing poetry, was used to create a compelling and, arguably, persuasive narrative. She positioned the conservationists' campaign as a 'David and Goliath' fight, despite a number of them receiving support from the federal government (Lloyd et al., 2017, p. 59).

A few weeks before her death in 2000, Wright wrote to activist Margaret Thorsborne, who worked with her on the Reef campaign, 'It is a long time since the reef was won, and it still surprises me that . . . it was won at all – if one can say that the present situation is a victory' (Wright, 2014, p. xviii). That *The Coral Battleground* was republished in 1996 and 2014 testifies to its significance as a record in what are unfortunately similar fights, particularly given that in June 2012, UNESCO threatened to add the Reef to the List of World Heritage in Danger unless the Queensland and federal government put actions in place to properly care for it. UNESCO were asked to consider adding the Barrier Reef to the in-danger listing over concerns around impact from coastal developmental on land adjacent to the Reef. The debate as to whether to list the Reef remains an ongoing issue.

Children's literature and the Great Barrier Reef

Children's literature makes up the largest genre of writing about the Reef listed in AustLit (that is, 14 per cent of the listings). Notably, the first children's book published in Australia, Charlotte Waring Atkinson's *A Mother's Offering to Her Children* (1841), featured the 1834 wreck of the 'Charles Eaton' near the Sir Charles Hardy Islands, cementing the Reef's danger in young Australian minds. In the AustLit database, the earliest listing in this genre is a short story published in 1879, 'Guinea Gold, or, The Great Barrier Reef'. In this boys' own adventure story, Charles H. Eden drew on his experiences of visiting the islands of the Reef 'for shooting purposes, to gather coral for limeburning, etc.' (Eden, 1879, p. 2). It features the story of Jack and Tom who, enroute from Papua New Guinea, become stranded on the Reef, then lose their ship and become separated in a cyclone. The Reef is infested with sharks, and the action is focused squarely on the humans.

By contrast, the most recent entry for children's literature in AustLit, Catherine Barr's 2022 *Let's Save the Great Barrier Reef: Why We Must Protect Our Planet*, illustrated by Jean Claude, is about how the Reef is valuable for a range of ecological, economic, and cultural reasons. It presents bright, full-page illustrations of underwater views of dugongs, coral, seaweed, turtles, whales, stingrays, clownfish, dolphins, stingrays, and sharks. Some illustrations feature both land and sea, showing plants and animals that utilise both terrains, such as turtles and mangroves. Each double page begins with the phrase 'Let's save the Great Barrier Reef' and continues with a paragraph of text explaining why – for example, 'because it's the BIGGEST living organism on our planet' ('it's the BIGGEST living organism' section) or 'because it keeps the Coral Sea clean' ('it keeps the Coral Sea clean' section). Smaller snatches of text are sprinkled amidst the illustrations, imparting information about different aspects of the Reef, such as the impact of climate change: 'In more acidic waters, corals struggle to build their strong rock-like skeletons and shellfish are less able to build hard, protective shells' ('it slows the waves and protects the shores' section). The effect of different sizes, shapes, and portions of texts points to the complexity of the Reef and the need for a range of modalities to communicate its intricacies.

Barr is a British author who drew upon advice from scientists Dr Karen Joyce, a remote sensing scientist at James Cook University, and Dr Tom Bridge and Associate Professor Alana Grech of the former Australia Research Council Centre of Excellence for Coral Reef Studies, James Cook University, signifying the continuing importance of communicating science about the Reef, which was emergent at the time of the 1960s campaigns. Science communication continues to be important, particularly given research indicating that literature about the environment engenders an appreciation of and empathy for the natural world (Babb et al., 2018, p. 716).

The work is also part of a trend in children's books which, although depicting an increasing number of representations of built environments rather than natural environments, also shows increasing sophistication in depictions of humans' interactions with nature (Babb et al., 2018, pp. 724–725). An analysis of 249 books from the Children's Book Council of Australia Picture Book of the Year Awards shortlists from 1955 to 2014 found that more recent works were more likely to depict ecosystems with 'a wider variety of life, multiple trophic levels, and complex interactions' (Babb et al., 2018, p. 725). This indicates an awareness of ecosystems and the work they do, reinforcing the concept that humans are never on their own, but are always knitted into a wider web of life – a critical concept to learn as humans and the Reef are impacted by a world of increasing heat.

Writing the Reef's future

Many of Barr's readers will grow up to be custodians of the Reef, but given the current trajectory of global heating, there may not be a Reef (at least as we now know it) for them to visit. Hughes et al. detail how cumulative bleaching in 1998, 2002, and 2016 created a footprint covering almost the entirety of the Great Barrier Reef Marine Park (2017, p. 373). Local management of coral reef fisheries and water quality offers little resistance in the face of these bleaching events, given that they are caused by global heating. Given this, communication to a wide audience about the wonders of the Reef and the threats to its existence is imperative.

Rooney suggests that the poetry Wright wrote during her activist years, 'even at its most prosaic, bitter or ugly, asks to be read as an exercise of defiance, flung out as a kind of hope or commitment against encroaching extinction' (2009, p. 20). Wright recognised the power of writing, whether poetic, political, or informative, to impart knowledge and galvanise action. Writing about the Reef, ranging from centuries-old stories from Aboriginal and Torres Strait Islander peoples to depictions of it as a liminal zone of beauty and restoration, to documents about its ecological significance and illustrated texts about its multifaceted uses, reflects a growing awareness of the Reef's indivisibility from humans' and nonhumans' survival.

I have not written the ending to my work-in-progress on the Reef and its creatures, as I am yet to decide whether it is a story of hope or despair. One element of the novel is clear, however: that without the rich colours, complex ecologies, and manifold diversity of life of the Great Barrier Reef, humans and nonhumans alike will be impoverished. While scientists at SeaSim are experimenting with breeding resilient corals, and writers are communicating the significance of the Reef, it behoves every human to work faster to contain global heating and protect one of the world's most incredible organisms.

Note

1 For more on the AustLit database, please visit https://www.austlit.edu.au/

References

AustLit. (1999). University of Queensland. http://www.austlit.edu.au

Australian Institute of Marine Science. (2023). 'National Sea Simulator.' https://www.aims.gov.au/about/facilities/national-sea-simulator

Babb, Y. M., McBurnie, J., & Miller, K. K. (2018). Tracking the environment in Australian children's literature: The Children's Book Council of Australia Picture Book of the Year Awards 1955–2014, *Environmental Education Research, 24*(5), 716–730, doi: 10.1080/13504622.2017.1326020

Banfield, E. J. (1908). *Confessions of a Beachcomber*. T. Fisher Unwin.

Barr, C., & Claude, J. (2022). *Let's Save the Great Barrier Reef: Why We Must Protect Our Planet*. Walker Books.

Bellwood, D., and Mallela, J. (2003). Biodiversity of the Great Barrier Reef. *Australian Academy of Science*. https://www.science.org.au/curious/great-barrier-reef

Bonyhady, T. (2005). Torn between art and activism. *Eureka Street*. https://www.eurekastreet.com.au/article/torn-between-art-and-activism

Braidotti, R. (2013). *The Posthuman*. Polity Press.

Cottrell, D. (1928). *The Singing Gold*. Hodder and Stoughton.

Devanny, J. (1944). *By Tropic Sea and Jungle*. Angus and Robertson.

Devanny, J. (1951). *Travels in North Queensland*. Jarrolds.

Eden, C. H. (1879). *Guinea Gold, or, the Great Barrier Reef*. Society for Promoting Christian Knowledge.

Grant, M. (1980). *Barrier Reef*. Hutchinson Australia.

Hughes, T., Kerry, J., Álvarez-Noriega, M., Álvarez-Romero, J., Anderson, K., Baird, A., Babcock, R., Beger, M., Bellwood, D., Berkelmans, R., Bridge, T., Butler, I., Byrne, M., Cantin, N., Comeau, S., Connolly, S., Cumming, G., Dalton, S., Diaz-Pulido, G., . . . & Wilson, S. (2017). Global warming and recurrent mass bleaching of corals. *Nature, 543*, 373–385. https://doi.org/10.1038/nature21707

James Cook University. (2022). Event: Writing at the Reef. https://www.jcu.edu.au/foundation-for-australian-literary-studies/news/event-2022-writing-at-the-reef-residency

Livett, K. (2015). The coral battleground (Book Review). *Southerly, 75*(1), 215–218.

Lloyd, R., Newlands, M., & Petray, T. (2017). Coral battleground? Re-examining the 'Save the Reef' campaign in 1960s Australia. *Environmental Sociology, 3*(1), 54–63, doi: 10.1080/23251042.2016.1259604

Love, R. (2000). *Reefscape: Reflections on the Great Barrier Reef*. Allen and Unwin.

Mairou, N. (2002). *The Dugong Meadow*. Trapdoor Press.

McCalman, I. (2013). *The Reef: A Passionate History*. Penguin Books.

McCalman, I. (2017). Linking the local and the global. What today's environmental humanities movement can learn from their predecessor's successful leadership of the 1965–1975 war to save the Great Barrier Reef. *Humanities, 6*(77), 1–12. https://doi.org/10.3390/h6040077

Miller, M. N. (2015). *Reef and Rainforest: An Aboriginal Voice through Art and Story*. RenBro Publishing.

Morrissey, D. (2008). *The Reef*. Pan Macmillan Australia.

Neimanis, A. (2017). *Bodies of Water: Posthuman Feminist Phenomenology*. Bloomsbury Academic. https://doi.org/10.5040/9781474275415

O'Connor, M. (1976). *Reef Poems*. University of Queensland Press.

Perchan, R. (2000). A golden meme of darkness: Darwinism, Robert Frost, and E. J. Banfield. *Antipodes, 14*(1), 41–45.

Perrett, L. (2010). *Sea Country Stewards: A Review of Traditional Owner Aspirations for the Great Barrier Reef*. Great Barrier Reef Marine Park Authority.

Pierce, P. (2004). Tropic days: Literature and art of North Queensland. *eTropic: Electronic Journal of Studies in the Tropics, 3*(1). https://doi.org/10.25120/etropic.3.1.2004.3443

Reid, F. (1954). *The Romance of the Great Barrier Reef*. Angus and Robertson.

Reid, N., & Nunn, P. (2015). Ancient Aboriginal stories preserve history of a rise in sea level. *The Conversation*, 13 January 2015. https://theconversation.com/ancient-aboriginal-stories-preserve-history-of-a-rise-in-sea-level-36010

Rooney, B. (2009). *Literary Activists: Writer-Intellectuals and Australian Public Life*. University of Queensland Press.

Ross, B. (1997). Different leaves from Dunk Island: The Banfields, Dorothy Cottrell and The Singing Gold. *LiNQ*, *24*(1), 56–70.
Stünzner, I. (2018). Traditional Owners capture Great Barrier Reef dreaming stories in new tourism venture. *ABC News*. https://www.abc.net.au/news/2018-04-28/traditional-owners-capture-great-barrier-reef-creation-stories/9707068
Taylor, C. (2004). Threshold to fulfillment: The Barrier Reef writings of E. J. Banfield and Jean Devanny. *Antipodes*, *18*(1), 18–23.
Torre, S. (2005). 'Reality' fiction and the Great Barrier Reef. *eTropic: Electronic Journal of Studies in the Tropics*, 4. https://doi.org/10.25120/etropic.4.0.2005.3440
Torre, S. (2008). Literary representations of the Great Barrier Reef. *eTropic: Electronic Journal of Studies in the Tropics*, 7. https://doi.org/10.25120/etropic.7.0.2008.3433
Wallace, J., & O'Carroll, J. (2016). Beyond subjectivity: The appearances of extinction in Judith Wright's 'Fourth Quarter' (1976). *Fusion Journal*, 10, 117–143.
Waring Atkinson, C. (1841). *A Mother's Offering to Her Children*. George Evans.
Weaver, R. (2015). Ecologies of the beachcomber in colonial Australian literature. *Journal of the Association for the Study of Australian Literature*, *15*(2). https://openjournals.library.sydney.edu.au/JASAL/article/view/9942
Wright, J. (1949). *Woman to Man*. Angus and Robertson.
Wright, J. (1955). *The Two Fires*. Angus and Robertson.
Wright, J. (1962). *Birds*. Angus and Robertson.
Wright, J. (1976). *Fourth Quarter*. Angus and Robertson.
Wright, J. (2014). *The Coral Battleground*. Spinifex Press.
Zeller, R. (2002). A Thoreau in paradise: E. J. Banfield's 'My Tropic Isle'. *The Concord Saunterer*, *10*, 134–145.
Zeller, R. (2004). E. J. Banfield as naturalist. *Queensland Review*, *11*(1), 17–25. https://doi.org/10.1017/S1321816600003536
Zeller, R. (2007). Tales of the Austral tropics: North Queensland in Australian literature. In C. A. Cranston & R. Zeller (Eds.), *The Littoral Zone: Australian Contexts and Their Writers*. Rodopi. https://doi.org/10.1163/9789401204514

PART 4

Beyond the anthropocentric blue

11

MOVING WATERS, MUDDY EDGES

Ibis in Brisbane

Gillian Paxton

One blustery morning, I was walking along the Eleanor Schonell Bridge, a tall pedestrian bridge that crosses the Brisbane River connecting the University of Queensland campus with the city's centre. From it, one can see far along the brown river. On this day, as I glanced down at the turbid water eddying dangerously at the change of tide, my attention was caught by a familiar shape floating amongst the clumps of leaves and plastic bottles pushed into the river by recent heavy rain. The shape was so at odds with its context that I stopped to make sure that it was indeed what I thought it was: the curved black beak of an Australian white ibis. A normally robust urban waterbird common in Brisbane, this ibis was clearly in trouble. Caught on debris, its body was water-logged and helpless beneath the water. With only its head breaking the surface, I assumed at first that the bird was dead. Then, to my horror, I saw it was making the smallest of movements. Every so often, it gently raised its beak above the water to take a breath then dropped it back down to rest. Profoundly moved by this persistence, I watched as the bird drowned and the current carried it out of view.

In cities, our relations with water and its capacity to engulf and overwhelm are largely managed for us. Vast infrastructures capture, purify, and reticulate water for our convenience, and buildings, roads, and pavements function as membranes that flush stormwater and rainwater off public and domestic spaces (Kaika, 2005; Walker et al., 2011). The massive effort it takes to protect us from water is often hidden, with underground pipes, sewers, and clever urban designs working to obfuscate the profound material privileges they bestow (Kaika, 2005). The rainwater that ambushed the ibis and pushed it into the brackish river was doing exactly what it is supposed to in a well-managed city: moving quickly out of the city via a series of gutters, culverts, and concrete stormwater drains before flowing, unnoticed, into the river and out to the sea. From the bridge, however, I glimpsed water acting

DOI: 10.4324/9781003365501-16

in ways that could not be fully reduced to this context, and was struck, in Jane Bennett's terms, by an 'eccentric outside' (2010, p. 45) to this civic performance. Caught by debris, surrounded by rubbish, and pushed into depths against which its body was powerless, I recognised the ibis as a ghastly consequence of the problematic 'blue' politics we enact as we seek to deny and forget the power of water in the city.

In this chapter I want to explore this 'blue' politics further, paying attention to the 'vibrant' (Bennett, 2010) but often troubling entanglements that form as Australian white ibis inhabit the watery spaces of the city. Beginning in the slow waters of Australia's largest river system, the Murray–Darling Basin, I illustrate how ibis have nimbly negotiated relations with the transient muddy edges created as floodwater saturates the earth. As the flows in this Basin have altered, ibis have proven equally adaptable, moving to the marginal wetlands of urban space and forming new attachments to the manufactured human food that circulates around the city. As they do, however, they disrupt our own efforts to negotiate our material entanglements, drawing our attention to flows we actively seek to ignore and muddying the rationalised edges of Brisbane's civic squares and public eating areas. Alerting us to vulnerabilities we would prefer to deny, we hasten to reassert our boundaries, deriding the ibis as a pest and 'trash animal' and subjecting it to control and violence. In highlighting this struggle, I invite readers to consider different ways of responding to the ibis' provocation, and what might be gained by celebrating the ibis' attentiveness to the city's lively flows and our own entanglements with the saturations of urban space.

Following flow

The ibis' story begins way west of Brisbane, in the northern sections of the Murray–Darling Basin. There, water moves and shapes the land in spatiotemporal configurations – or what Jones (2011) elegantly refers to as 'rhythmpatterns' – which are the most variable and intermittent in the world. Prolonged dry periods reduce the flat catchment to desiccated plains, dusty riverbeds, and a spattering of permanent waterholes. When the rain comes, infrequently but in great volumes, it seeps through the landscape, pooling in low-lying areas and swelling over shallow riverbanks to the floodplains beyond. There is little to channel this water away, so it sits shallow on the dry land, softening it into lush marshes, mudflats, and temporary lakes (Kingsford & Norman, 2002). This water is rich with organic sediment, minerals, and nutrients, and as it soaks into the dry earth, tiny single-celled life lying dormant in the dirt becomes plump and animated. As it does, larger invertebrates in turn multiply and swarm.

For millennia, Australian white ibis, as well as other Australian waterbirds, have cultivated the 'boom and bust' (Robin & Joseph, 2009) seasonality of the floodplains and the temporary, transitional ecologies created at the muddy edges of flooded land. These relations have deeply affected their lives and shaped their bodies, habits, and skills. Strong flyers and highly mobile, Australian white ibis have developed a canny awareness of, and attention to, inland flows. During dry times, ibis typically stay in small colonies around permanent wetlands or move into coastal areas (Kingsford et al., 2010). It is not clear how they know, but when the rains come inland, these birds abandon these colonies and travel hundreds of kilometres to the floodplains. There, in massive flocks, they stalk the mud, probing for unseen critters with sensitive and specialised beaks. They readily translate this glut of biomass into new life, with females laying more eggs, and more frequent clutches, during times of flood. However, just as water provides life, it can also overpower. As much as the ibis has been shaped in its cultivation of water, it has been equally shaped by its need to negotiate its intermittent flows and hold itself back from its material capacities. Ibis bodies, for example, keep them equal to the intermingled material worlds they inhabit, with long legs to keep feathered bodies above the water, splayed feet to find purchase on the softest ground, and featherless heads to dip without becoming sodden. And as much as they attend to flows and flourish at the floodplain's muddy edges, they do not become too obligated to the floods. When the land begins to shrink and harden, ibis are quick to withdraw their connections to the floodplains, take flight, and seek other watery opportunities (Carrick, 1962; Kingsford et al., 2010).

When European settlers arrived in the Murray–Darling and laid claim to its floodplains for sheep and cattle, they were less adept than the ibis at negotiating the unpredictable material flows of water through the landscape. While they welcomed periodic inundations for the eruption of plant life they catalysed, they also feared for newly established townships and agricultural communities (O'Gorman, 2012; O'Gorman et al., 2016). In the early days of agricultural expansion, settlers sought to assert new cultural boundaries of 'spatially constituted normality' (Walker et al., 2011, p. 2306), defining floods as climatic anomalies and threats to be managed in the interests of colonial establishment. Managing floods demanded acts of human ingenuity and the construction of levies and weirs, and, in romantic narratives of the time, pioneering spirit, endurance, and tenacity. For a time, ibis were enrolled as symbolic allies to these endeavours, featuring in sentimental pastoral art and scientific journals as the 'Farmers' Friend' (Pennycook, 1930). Naturalists described them as 'the most useful birds Australia possesses' (le Souëf, 1917, p. 95) for their readiness to cultivate the soft edges of flooded pastures for insect pests. As agricultural development shifted from colonial establishment to intensified production and national progress, new industries such as rice and cotton needed more secure water. By the middle of the twentieth century, large-scale irrigation projects were diverting enormous volumes of Murray–Darling water into storages and dams (O'Gorman, 2012; Robin & Joseph, 2009). As these flows were captured and productivity was controlled, little care was given to those that had a stake in the Murray–Darling's ephemerality and its 'boom and bust' seasonality. As greater volumes of water were diverted, flooding became less frequent and the flows that did manage to reach the floodplains were reduced, bringing a range of impacts to rangeland pastoralists and wetland ecologies (CSIRO, 2008). As farmers and environmentalists lobbied for greater recognition of floodplain interests and a more equitable allocation of water through the Murray–Darling Basin, they held up the fragile symbiosis between floodplains and inland waterbirds – including the Australian white ibis – as an example of how much was at risk (Carrick, 1959).

Downtown waterbirds

While the fears of environmentalists have mostly been realised, they underestimated the dexterity of the connections that Australian white ibis form with the inland floods. As the volumes of flows reaching the floodplains declined and the regularity of floods dropped, the appropriation of water in the Murray–Darling has become one of the most significant environmental issues in Australia. The ibis, however, did not go down with the ship. Instead, from the 1970s onwards, greater numbers of ibis became observed in cities along the eastern coast of Australia (Ross, 2004). While this transition to urban life surprised many, it was not inconceivable. Cities still offer an array of ecological opportunities for wildlife, including waterbirds. Situated near the mouth of the Brisbane River, criss-crossed by creeks and springs, and periodically flooding, Brisbane is shaped by its own watery 'rhythmpatterns' and while it is frequently subject to human control, water can still contravene the script set out for it. While much of Brisbane's stormwater is diverted away from residences and public areas, some still accumulates in the city's low-lying areas, making them unsuitable for human development. These sodden spaces become non-descript wetlands amid the city's industrial zones and road networks. Every evening they are transformed into raucous habitat, as thousands of waterbirds – including cattle egrets, herons, magpie geese, and Australian white ibis – fly in after a day's foraging and settle for the night into the surrounding scrub. As they dwell in the city's wet places, this community of waterbirds is testimony to a city emergent in and enlivened by water flowing 'against the grain of expert designs and blueprints' (Hinchliffe & Whatmore, 2006, p. 134).

However, ibis are not simply eking out an existence in the overlooked wet spaces of the city. As the ibis has become urban, it has expanded its ecological entanglements to include other material circulations entwined in, and shaped by, the city's social, political, and economic practices (Kaika, 2005; Swyngedouw, 2004). This is most evident during the day, when ibis leave their communal roosts to forage. While most of the city's waterbirds travel to Brisbane's creeks, mudflats, estuaries, and grasslands, ibis travel to civic squares, outdoor shopping malls, school playgrounds, university campuses, and municipal waste processing facilities (Murray, 2005; Ross, 2004). Instead of foraging the invertebrates proliferating in the sodden ground, they spend the day searching for and ingesting the nutrients that accumulate as city workers, shoppers, and tourists purchase, consume, and discard manufactured food in the city's commercial and civic spaces.

For an animal so well adapted to the ephemeral flows of the floodplains, expanding their connections in this way is no mean feat. To achieve it, the ibis has had to experiment with new ways of cultivating and negotiating urban flows, spaces, and infrastructure. It is more than capable of moving from the soft surfaces of the wetlands to the dry, hard spaces where humans assemble and consume food, and it appears equally at ease poking its head into garbage bins as it does probing the mud of the floodplains. It exhibits an exceptional dietary flexibility and has developed the skills to manipulate manufactured human food like pizza, chips, and hamburgers into its gullet. To add to this list of urban accomplishments, the ibis has proven itself adept not only at tolerating the presence of large numbers of humans but actively attuning to the flows and rhythms of their food consumption. Tracking studies in Brisbane show ibis have an idiosyncratic knowledge of the places in the urban landscape where food is publicly consumed and discarded and align their foraging routines to the times when humans eat. They will arrive at school playgrounds, for example, during lunch and recess times and depart soon after the bell rings (Murray, 2005). When office workers, tourists, and students congregate at Brisbane's outdoor eating areas and fast-food outlets, ibis will stalk amongst the benches and tables watching for food left exposed. They will readily steal a chip from a child's hand or ransack unattended tables.

FIGURE 11.1 Australian white ibis at the University of Queensland, Brisbane.

The ibis' readiness to form connections with human food in the city has yielded it tangible ecological benefits. By tapping into the overflows and excesses of food consumption in the city, urban ibis ingest more calories within a shorter amount of time and travel smaller distances to secure sources of nourishment. As a result, urban ibis populations have grown exponentially in comparison to populations outside the city (Murray, 2005; Ross, 2004). By cultivating the flows of the city, ibis have secured themselves a future, as the water moving through their traditional floodplain habitat has been increasingly moderated to serve the interests of post-settlement agriculture. Ibis do not simply exist in cities, but flourish in them due to their willingness to experiment with new entanglements, new materiality, and new ways of being. In connection with humans and the flows of food circulating through the city, ibis live and breed as in permanent flood. As they do, ibis have become familiar urban animals, embedded in the rhythms and flows of public life in the city. As they walk unabashed through built-up boulevards, civic squares, and shopping malls, ibis influence and enliven the fabric of the city and the ways it is experienced by humans. For this, the ibis becomes drawn into human efforts to supress, control, and order the lively flows of the city (Kaika, 2005).

Provocation and control

In *Ethics of Waste*, Hawkins (2006) describes how waste makes a material claim on us that is deeply emotional, involving disgust, irritation, guilt, and shame. Like the water flowing through cities, our vulnerability to waste is managed through urban processes and infrastructures to remove it and keep it out of sight (Hird, 2013; Kaika, 2005). Mundane acts of waste disposal are part of this cultural performance, but there are moments – such as when the bin overflows – when tensions emerge in this order. We are then provoked into noticing and 'feeling waste' (Hawkins, 2006, p. 1), and this signals 'a kind of failure' (Hawkins, 2006, p. 1) in the domestic practices of civilised modernity. Public eating spaces like the Central Café Precinct at Brisbane's Southbank Parklands are designed to facilitate the efficient and economic consumption and disposal of food, with proprietors serving up cheap, calorie-rich food to consumers, who eat quickly, vacate their table, dump their scraps, and make room for the next customers to eat. However, similar tensions to those described by Hawkins (2006) can be felt as ibis stalk around these spaces, rummaging through bins and grabbing unattended food. Diners must work to protect their food from ibis, using their hands and feet to shoo ibis away, and signs warn diners to 'Be Cautious' of the birds. In what should be an unnoteworthy episode in a family day out, parents tell their children to hurry up and eat, and when ibis do manage to steal food from a table, there is a flurry of harsh words and feathers and a collective sense of embarrassment, ignominy, and irritation.

FIGURE 11.2 Australian white ibis at Southbank Parklands, Brisbane.

Like the stormwater that ambushed the ibis in the Brisbane River, excess food is meant to be channelled out of the public spaces of Brisbane via a range of disposal practices and technologies, and enter massive waste management infrastructures to be processed and, eventually, forgotten (Hird, 2013; Kaika, 2005). However, ibis recognise the nutritional bounty contained in this flow and readily swarm to take advantage of it. As marauding packs of these waterbirds cultivate urban food waste, much like they would the invertebrates that flourish in recently flooded pastures, this performance fails. Like an overflowing bin, excess food cannot be ignored and this awareness is uncomfortable for humans. With the presence of the ibis, the boundaries enacted in the dry modern eating spaces of the city become less clear. These places no longer feel like shining monuments to humanity's technological prowess and advanced, enlightened civilisation. They become saturated, edge ecologies that highlight our embeddedness within, and vulnerability to, a multitude of non-human interests and fluid material forces.

The discomfort that ibis provoke as they disrupt our careful performances around public food and make us acknowledge the material excesses we would prefer to ignore, has made the constituency of the ibis in cities tenuous. This is reflected in the narratives we use to construct the ibis. Despite being native to Australia and formally protected under conservation laws, urban ibis are frequently labelled as urban nuisances or 'pests' that threaten civic amenities and 'conflict with human desires of what nature in the city *ought* to be' (McKiernan & Instone, 2016, p. 9, original emphasis). These narratives have very real implications for ibis, acting to make them a 'killable' (Haraway, 2008) aspect of the urban environment. A relational practice distinct from the actual act of killing, making things killable is a political practice designed to not take the act of killing seriously and involves processes whereby the livingness of something and the richness of its history are erased to hide what is lost through its killing (Haraway, 2008). As they render the ibis unwelcome in urban environments, pest narratives legitimise and sanction council-led programmes to deter ibis from the city. Implemented by environmental management companies up and down the eastern coast of Australia, these are not aimed at complete extermination, but use both lethal and ostensibly 'non-lethal' measures to curb ibis fecundity and sever their ecological connections with the flows of food waste through the city. At regular intervals through the year, ibis managers enter key ibis roosting sites to destroy nests and eggs, conduct coordinated ibis deterrence activities at urban garbage dumps, and deter ibis from public spaces using gas guns, dogs, and noise deterrents. Garbage bins across the city are carefully designed to prevent ibis from having easy access to the bounty within, and signs discourage users of public spaces from sharing food with ibis.

By defining ibis as problems to be managed in the name of civic amenity, pest narratives facilitate the ibis' subjection to biopolitical programmes of control. Other narratives go further. In online discussions (see for example, Stephens & Lawrie, 2015), urban ibis are described as 'vermin', 'the equivalent of flying rats', and 'blights' on the city. These words marginalise and degrade the urban nature in which the ibis participates. Because they form connections to flows of urban waste, ibis are not just considered animals out of place, but become part of a community that Nagy and Johnson (2013) refer to as 'trash animals': animals that are worthless, useless, and disposable. The discourses that construct these urban animals extend beyond the rationalised killability that underpins urban environmental management and include more vitriolic narratives that scorn and devalue the ibis' presence in the city (Nagy & Johnson, 2013). These justify cruelty against the ibis as acts of retribution. For example, despite only being a fraction of the size of a human and engaging in neither territorial nor aggressive behaviour, ibis are sometimes attributed with having thuggish intent and are accused of 'attacking' humans as they approach them for food. In 2011, to the horror of onlookers, a young law student eating his lunch at one of the outdoor tables in the Southbank parklands repeatedly kicked and stamped on an ibis that had attempted to steal his food, injuring the bird to such an extent that the police were called and the bird was subsequently euthanised. As he pleaded guilty at the Brisbane Magistrates Court to one count of animal cruelty, his justification for his actions was his need to 'defend' himself from the ibis (Lill, 2011). As well as being constructed as thugs and threats to human security, ibis are also ridiculed and mocked online for the connections they form in urban space and materiality. On the Ibis Appreciation and Recognition Facebook page, for example, ibis are derided as if they were hobos or vagrants. They are described as 'bin chickens' addicted to 'bin juice'. As a response to a real incident in Brisbane's Queen Street Mall in January 2017 (Courier Mail, 2017) in which an ibis was strangled by a man and its body used to intimidate bystanders, the Facebook page organised a mock 'candlelight chip eating vigil' in the ibis' honour. Devaluing the ibis' life in the city is an assertion of sovereignty over urban space and an active denial of our entanglements.

Conclusion

On that day on the bridge, I struggled to articulate how much my encounter with the ibis in the Brisbane River affected me. I knew it was transformative, like the one described by Jane Bennett (2010) when, one morning, she observed in the grate of a drain near Chesapeake Bay a man's plastic glove, a dead rat, a bottle cap, a mat of oak pollen, and a stick of wood. Struck by the vibrancy of this 'contingent tableau' (Bennett, 2010, p. 5), Bennett was able to recognise not simply a bunch of unrelated, inert objects, but a vivid assemblage of things that could not be reduced to the contexts in which humans had set them and were 'never entirely exhausted by their semiotics' (2010, p. 5). In the modernist, human-transcendent politics that infiltrates the stories we tell about cities, it is often assumed that they are rendered separate and autonomous from influences of nature and that we are their sole creators and beneficiaries (Hinchliffe, 2007; Kaika, 2005). In that swirling messy assemblage of water and life below me on the Eleanor Schonell Bridge, I became struck by something more interesting: a city enacted at the intersections of many material and biotic flows and enlivened by more than human histories, practices, and intentions. Humans may indeed be system-defining actors in processes of becoming urban, but other things also exert influence and pull the urban fabric in manifold directions. In that moment, the ibis became a provocation to recognise these Others (Haraway, 2008) and 'the extraordinary that lives among the familiar and the everyday' (Bennett, 2010, p. 4).

This chapter has attempted to pay careful attention to the realities that emerge as water entangles with life in urban space. It has shown the ibis attuning to the ephemeral flows that soak and seep through the inland floodplains, nimbly 'holding its shape' (Hinchliffe, 2007) against water's material capacities and cultivating the nutritional eruptions it catalyses. It has also shown how, in the face of greater control and regulation of ephemeral inland floods, the Australian white ibis has performed a surprising ecological transition to the city, joining a thriving community of urban waterbirds in the otherwise overlooked muddy edges of the Brisbane River. The ibis has expanded its connections to new material flows, directly cultivating humans for food and becoming a ubiquitous presence in the public spaces of Brisbane. As it does, it is a bawdy contradiction to any human-transcendent notion that water's exuberance can be erased from urban landscapes, and points instead to the muddy edges of the city created in the material flows of human food consumption.

The ibis in the Brisbane River is not only a provocation to attend to the city as a muddy, lively, entangled place. It is also a reminder of why, when Haraway tells us that 'to be one is always to become with many' (2008, p. 4, original emphasis), she is careful to point out that that enacting reality in assemblage is unavoidably fraught. Giving to ontological relations also requires entities to hold something in reserve so they do not become overwhelmed by the relations and realities that emerge. This is a critical point for thinking about entanglements in the 'blue'. Yes, our lives are unavoidably entwined with flows of water, but this does not mean that the relationship is easy or pleasant. In fact, we must constantly find ways to hold ourselves back from water. In modern cities, we do this by developing vast systems to control our encounters with water and minimise our vulnerability to it (Kaika, 2005; Walker et al., 2011). But as this chapter has shown, there is collateral damage to this performance, as the ibis disrupts our delusions of separation and control as it forages Brisbane's civic spaces and outdoor eating areas. By provoking us to feel the watery realities we would rather ignore, we typically double-down on our effort to divide and deny: ibis become the targets of disdain and hatred and are subject to official and unofficial forms of violence, exclusion, persecution, and control.

Acknowledging our entanglements in the blue means we must critically interrogate the ways we exert influence within these entanglements. We must ask ourselves: might there be more honourable ways to manage our connections to the material flows of urban spaces? For Despret and Meuret (2016), a true politics of attention is never simply about taking others into account. As we acknowledge the myriad other forces, flows, and lives that comprise the city, our ethical capabilities and obligations to them must follow: 'ecology and cosmology knotted in a common story' (Despret & Meuret, 2016, p. 25). As humans have played an instrumental role in the creation of cities and their unequal and fraught urban relationships, we must allow these situations the power to obligate us 'without guarantees' (Despret & Meuret, 2016, p. 27), and without seeking to control all the outcomes. In this sense, the ibis in the Brisbane River is a provocation that alternatives to human sovereignty, control, and transcendence are necessary and will require

> learning to hold possibilities open, learning attentiveness to the infinite ways of being affected and of affecting, where no one may know ahead of time the affects one is capable of, or the kinds of forces and entities that will constitute landscapes and worlds with us.
>
> *(Despret & Meuret, 2016, p. 35)*

This means not only acknowledging the more-than-human forces of cities but being willing to treat the 'risky attachments' (Instone, 2015) that form in these urban entanglements as catalysts for new possibilities and solutions rather than threats to eradicate.

With its keen attention to overlooked urban spaces and an ability to revel in the ebbs and flows of the city's abject material excesses, the ibis has much to teach us about grappling with risky attachments. In Brisbane, I see glimmers of this in the graffiti around the city which celebrates the ibis in urban life as an enrichment rather than a nuisance, and in small human acts of transgression that challenge the divisions and hierarchies of urban spaces to give voice to often overlooked and degraded constituents. In one of these, students at the University of Queensland scavenge the university's cafés and bars for discarded food. This practice, which they refer to as 'ibissing', intentionally highlights the wealth that becomes discarded in everyday urban practice and celebrates the wisdom and knowledge of those that can recognise it. Another is illustrated in the image in Figure 11.3, which shows the logo of the Brisbane Free University.

FIGURE 11.3 The Brisbane Free University logo (reproduced with permission from artist Anna Carlson).

This informal civic institution seeks to 're-create university-like spaces of learning according to their own radical visions of social justice' (Thompsett, 2016), bringing academic and political discussion onto city streets. Fortnightly lectures focus on a variety of urban and political issues and are held in an open access carpark in the inner-city suburb of West End. Run by volunteer academics, activists, artists, the homeless, and others, the organisers are committed to operating without cash and never compromise on providing free entry (Thompsett, 2016). They openly attribute credit to the ibis for their inspiration and have chosen it as the organisation's logo in honour of 'its antagonistic scavenger spirit' (McMillen, 2015, p. 30) and its dogged attentiveness to the unacknowledged and 'free' flows of value and richness in the city. In these ways, the ibis in Brisbane is already helping us to experiment with our vulnerabilities to the material flows and excesses of urban life, and work towards more honest, generous, and ethical 'blue' entanglements.

References

Bennett, J. (2010). *Vibrant Matter: A Political Ecology of Things*. Duke University Press.
Carrick, R. (1959). The food and feeding habitats of the Straw-necked Ibis, *Threskiornis spinicollis* (Jameson), and the White Ibis, *T. molucca* (Cuvier) in Australia. *CSIRO Wildlife Research, 4*, 69–92.
Carrick, R. (1962). Breeding, movements and conservation of ibises (*Threshkiornithidae*) in Australia. *CSIRO Wildlife Research, 7*, 71–88.
Courier Mail. (2017, 10 February). Accused ibis killer to face Brisbane court.https://www.couriermail.com.au/news/queensland/crime-and-justice/accused-ibis-killer-to-face-brisbane-court/news-story/ee0ca61d77d849eb4a7025fbfea11125
CSIRO. (2008). *Water availability in the Condamine–Balonne*. A report to the Australian Government from the CSIRO Murray–Darling Basin sustainable yields project.
Despret, V., & Meuret, M. (2016). Cosmoecological sheep and the arts of living on a damaged planet. *Environmental Humanities, 8*(1), 24–36. https://doi.org/10.1215/22011919-3527704
Haraway, D. (2008). *When Species Meet*. University of Minnesota Press.
Hawkins, G. (2006). *Ethics of Waste: How We Relate to Rubbish*. UNSW Press.
Hinchliffe, S. (2007). *Geographies of Nature*. SAGE Publications Ltd.
Hinchliffe, S., & Whatmore, S. J. (2006). Living cities: Towards a politics of conviviality. *Science as Culture, 15*(2), 123–138. https://doi.org/10.1080/09505430600707988
Hird, M. J. (2013). Waste, landfills, and an environmental ethic of vulnerability. *Ethics and the Environment, 18*(1), 105–124. https://doi.org/10.2979/ethicsenviro.18.1.105
Instone, L. (2015). Risking attachment in the Anthropocene. In K. Gibson, D. Rose, & R. Fincher (Eds.), *Manifesto for Living in the Anthropocene* (pp. 29–36). Punctum Books.
Jones, O. (2011). Lunar–solar rhythmpatterns: Towards the material cultures of tides. *Environment and Planning A, 43*(10), 2285–2303. https://doi.org/10.1068/A4468
Kaika, M. (2005). *City of Flows: Modernity, Nature and the City*. Routledge.
Kingsford, R., & Norman, F. (2002). Australian waterbirds – products of the continent's ecology. *Emu, 102*, 47–69.
Kingsford, R., Roshier, D., & Porter, J. (2010). Australian waterbirds – time space travelers in dynamic desert landscapes. *Marine and Freshwater Research, 61*, 875–884.
le Souëf, W. (1917). Australian ibises. *Emu, 17*, 94–95.
Lill, J. (2011, 16 March). 'Law student Andrew Quay Wee Meng stomped and killed ibis which stalked him for sandwich at South Bank'. *The Courier-Mail*. https://www.couriermail.com.au/news/queensland/animal-cruelty-conviction-after-law-student-andrew-quay-wee-meng-stomps-on-ibis/news-story/0a9edb13af8e68a8d60d42231455b11b
McKiernan, S., & Instone, L. (2016). From pest to partner: Rethinking the Australian White Ibis in the more-than-human city. *Cultural Geographies, 23*(3), 475–494.
McMillen, A. (2015, November). Freedom thinkers: Brisbane Free University. *The Courier-Mail*.
Murray, N. (2005). Diurnal movements, behavior and foraging ecology of the Australian White Ibis (*Threskiornis Molucca*) in an urban landscape [Honours thesis]. Griffith University.
Nagy, K., & Johnson, P. (2013). *Trash Animals: How We Live with Nature's Filthy, Feral, Invasive, and Unwanted Species*. University of Minnesota Press.
O'Gorman, E. (2012). Local knowledge and the state: The 1990 floods in Cunnamulla, Queensland, Australia. *Environmental History, 17*(3), 512–546. https://doi:10.1093/envhis/ems045
O'Gorman, E., Beattie, J., & Henry, M. (2016). Histories of climate, science, and colonization in Australia and New Zealand, 1800–1945. *Wiley Interdisciplinary Reviews: Climate Change, 7*(6), 893–909. https://doi.org/10.1002/WCC.426

Pennycook, W. (1930, 30 August). The farmers' friend: The Ibis and its rookeries. *The Argus.*

Robin, L., & Joseph, L. (2009). The boom and bust desert world: A bird's eye view. In L. Robin, R. Heinsohn, & L. Joseph (Eds.), *Boom and Bust: Bird Stories for a Dry Country* (pp. 7–34). CSIRO Publishing.

Ross, G. (2004). Ibis in urban Sydney: A gift from Ra or a Pharoah's curse? In D. Lunney & S. Burgin (Eds.), *Urban Wildlife: More than Meets the Eye.* Royal Zoological Society of New South Wales.

Stephens, K., & Lawrie, J. (2015, 13 February). South Bank ibis deterrent proves music to birds' ears. *Brisbane Times.*

Swyngedouw, E. (2004). Metabolic urbanization: The making of cyborg cities. In N. Heynen, M. Kaika, & E. Swyngedouw (Eds.), *In the Nature of Cities.* Routledge.

Thompsett, F. (2016, 5 September). *Learning by doing by learning: Reflections on scholar-activism with the Brisbane Free University*, National Tertiary Education Union. https://archive.nteu.org.au/adfa/article/Learning-by-doing-by-learning%3A-Reflections-on-scholar-activism-with-the-Brisbane-Free-University-%28AUR-58-02%29-18961

Walker, G., Whittle, R., Medd, W., & Walker, M. (2011). Assembling the flood: Producing spaces of bad water in the city of Hull. *Environment and Planning A: Economy and Space, 43*(10), 2304–2320. https://doi.org/10.1068/a43253

12

A WHALE OF A JOURNEY

On the connectivity between pygmy blue whales in Indonesia, Australia, and beyond

Putu Liza Kusuma Mustika

It was another hot morning in November 2021. We have been at sea for more than ten days, cruising the edges of the Savu Sea, Indonesia, on the liveaboard ship, the *Pindito*, doing an ecological survey on the cetaceans in the area: whales and dolphins. In addition to estimating the relative abundance of the cetaceans, we also needed to deploy three satellite tags on pygmy blue whales (*Balaenoptera musculus brevicauda*) that regularly roam the Savu Sea as part of their annual journey between Banda Sea (Indonesia) and Australia. However, as we started the project in mid-2020 just as the COVID-19 pandemic occurred, it had been very hard to bring aboard an overseas researcher to help us tag the whales. As an Indonesian passport holder who resides in Australia, I had to undergo five days of quarantine in Jakarta before I could fly to Kupang, the closest harbour town in West Timor from which our liveaboard vessel departed for this project. As a result, our team had to do the tagging on our own, after consulting with various researchers.

This preparation led to ten days at sea and at least half a dozen attempts to deploy the satellite tags to pygmy blue and sperm whales – none of them successful. Finally, we encountered a large pygmy blue whale, cruising silently from the direction of Ombai Strait, which is the gateway from Banda Sea to Savu Sea. In fifteen minutes we gathered our kits, jumped to our zodiac boat, and zoomed out into the water. We had reviewed our previous drone video footage to realise that we had to really slow down as we approached the whale, otherwise we knew it would flee.

At the helm of our small zodiac, Oom Apé, the formidable First Officer of the *Pindito*, slowed the boat down, as the whale's torpedo shape was within sight. Jaya Ratha and Rusydi – two scientists from Reef Check Indonesia and Kupang Muhammadiyah University, respectively – who had been handling the satellite tag

DOI: 10.4324/9781003365501-17

deployment for this project, then prepped the tag launcher.[1] As Rusydi calmed himself, he aimed, shot a Wildlife Computers SPLASH10–373 tag using the Restech Norway's ARTS™ whale tag launcher, and after many failed attempts over the past ten days, he finally managed to securely plant one satellite tag into the whale's upper dorsal area.

There was a split second of silence before the whole team aboard the zodiac erupted in joy. We asked the remaining team at the *Pindito* to observe the whale's movements as we turned the zodiac around and hurriedly returned to the ship. On deck, we observed the whale until we were certain that it was not too stressed from the tagging procedure and that it kept going. We were buoyant for the remainder of the day, and I sincerely hoped – and still hope – that the whale was okay too.

I won't lie here, tagging is an intrusive activity. I used to think that tagging was cool . . . until I did it myself. I've changed my opinion since we approached the first whale during our Savu Sea survey in November 2021. Tagging is not a glorified business. It is not cool. It can be very intrusive, and it can put a lot of stress on the animals. That's why upholding animal ethics procedures is vital, and why scientists need to understand that every ounce of data derived from tagging is obtained by putting a lot of stress on the animals. We need to respect them, and one way of doing it (other than ensuring that the stress during the deployment process is minimal) is to fight for their protection.

We tagged another whale two days later, this time, a sperm whale (*Physeter macrocephalus*). Then, on 19 November 2021, the day before we ended the survey, Rusydi and Jaya Ratha were able to deploy the last tag onto the last pygmy blue whale of our trip.

The sperm whale tag only lasted for seven days before we lost its signal, while it was seemingly still roaming the western waters of Timor Island. The two pygmy blue whales, however, transmitted their signals for the next two to six months. And thus, their stories unfolded before our eyes.

The threats humankind poses to blue whales

The pygmy blue whale is but one sub-species of blue whales currently inhabiting the Earth's oceans. Scholars have identified four sub-species of blue whales worldwide: the Northern blue whale (*Balaenoptera musculus musculus*), the Antarctic blue whale (*Balaenoptera musculus intermedia*), the pygmy blue whale (*Balaenoptera musculus brevicauda*), and the Northern Indian Ocean blue whale (*Balaenoptera musculus indica*) (Branch et al., 2007). A fifth, presently unnamed, sub-species often observed off Chile was later suggested as an addition to this list (Leslie et al., 2020). Just like many other great whales, blue whales are often sighted in waters with dynamic oceanographic processes (such as upwellings) as well as areas with high density of phytoplankton, which is a main food source (Branch et al., 2007). By virtue of their very large size (up to 30 metres long), blue whales in general were one of the main targets of whaling. Commercial whaling, which started in the eighteenth century, hit this population badly, but the species has been experiencing a slow recovery thanks to the whaling moratorium enacted by the International Whaling Commission in 1986.

Whaling is not the only threat this species is facing (nor the only threat facing other whale and dolphin species, for that matter). The fisheries industry, marine tourism, marine traffic, offshore mining activities, and marine debris (including the ubiquitous plastic) are threatening their lives in this modern Anthropocene (see also Oppermann, 2023, pp. 5, 31). I've worked with enough cetacean populations in Indonesia to know that they are indeed threatened by humans. I've spent months with the traditional whale hunters of Lamalera (Lembata Island, East Nusa Tenggara) to understand their perspectives on hunting. I have spent at least two years in Lovina, north Bali, to observe and analyse the dolphin-watching tourism there to understand both the ecological side (the tour boats are adversely impacting the dolphins [Mustika et al., 2015]) and the human dimension side (these boatmen need a livelihood, thus they will keep taking tourists to see the dolphins [Mustika et al., 2013]). I've seen and read about enough cetaceans killed through their unfortunate interactions with the fisheries industry, whether as bycatch (unintended catch) or due to entanglement with fishing gears (Cassoff et al., 2011; Reeves et al., 2013). Solving human–animal conflicts, particularly when human livelihood is at stake, is not easy. The nuances are more complicated in developing countries like Indonesia (the fishermen whose nets had caught the animals could be very remorseful, yet they still had to fish for a living), but it does not mean that this issue is easy to solve in developed countries such as Australia.

That is why it is so important for us to understand where the Savu Sea pygmy blue whales are going, what marine habitats they are using, and how long they are in a particular patch of marine region, so that we can better protect them.

A journey of two whales

I return now to our duo pygmy blue whales. For simplicity, I'll call the first whale we tagged on 16 November 2021 Whale One, and the second whale we tagged on 19 November 2021 Whale Two. We could not obtain the genetic material from those whales despite our efforts to collect skin samples using darts and e-DNA. Since their sex is unknown, I will just use the pronoun 'it' when talking about each of them.

I was not expecting anything from the whales' tracking data, other than confirming past studies. We anticipated that the whales would cruise the Western Australian waters before the transmissions ended. My crew were happy enough that we were able to deploy three satellite tags to two pygmy blue whales and one sperm whale.

But in late December 2021, I realised that Whale Two remained persistent in its travel towards the southwest waters of Australia . . . and beyond. That was unexpected. The expected trajectory was that the whale would cruise the Western Australian waters before turning east, moving along the Great Australian Bight before, perhaps, exploring the southern waters of Australia. I, along with my co-researchers Jaya Ratha and Edy Setyawan, was not expecting the whale to travel beyond Australia's Economic Exclusive Zone (EEZ) into the sub-Antarctic waters.

I then checked Whale One, which took the expected route, and by this time, it was still cruising nominally along the Western Australian waters, veering close to the coast. It might turn east to cruise the southern waters of Western Australia, I thought, and true enough, it did. That was to be expected. Nothing special to report back. Until, in early January, Whale One also veered off the expected track (Figure 12.1). Around 500 kilometres south of Esperance (Western Australia), it actually turned southward and started exploring the waters between the Subtropical Front and the Sub-Antarctic Front, south of Australia's EEZ. That was definitely unexpected.

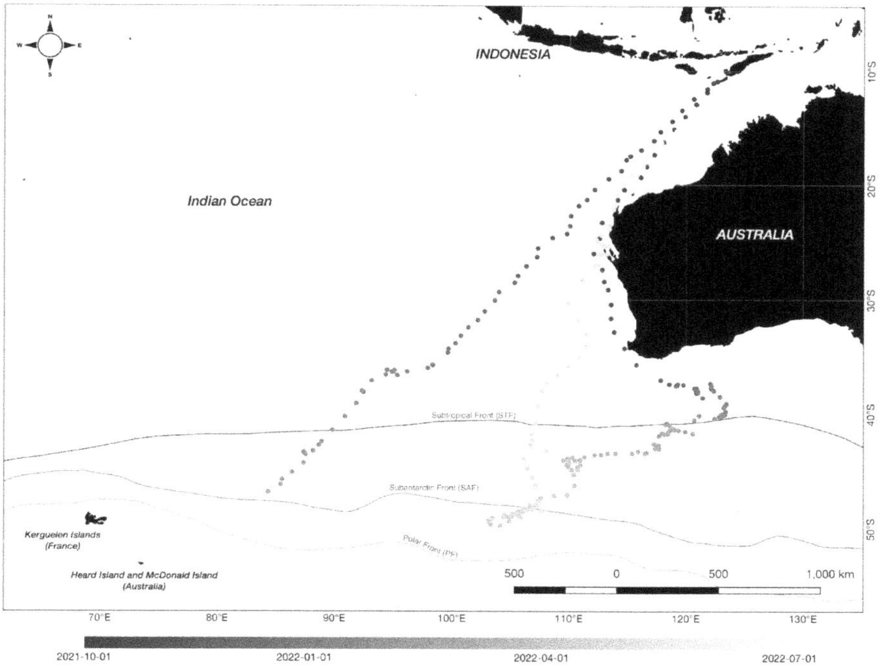

FIGURE 12.1 The tracks of two pygmy blue whales from Savu Sea (Indonesia) to the waters of Western Australia and the Sub-Antarctic.

Source: Map by Edy Setyawan.

For readers unfamiliar with oceanography, ocean fronts are very important features in understanding the ocean and its associated marine life. Fronts create upwelling (ocean water from the depth brought up to the surface, or near-surface of the ocean) by virtue of massive differences in sea water temperature, hence, they are ideal hanging places for marine life, including great whales. The fact that Whale One chose to hang out in the waters between the Subtropical Front and the Sub-Antarctic Front was not a surprise for me, because ocean zones around any fronts are abundant with their prey. The fact that it roamed the southern waters of Australia beyond the EEZ was the surprising factor here, for other scientists (Double et al., 2014; Möller et al., 2020; Thums et al., 2022) have never observed pygmy blue whales beyond Australia's southern EEZ.

And the surprises did not stop there. The other whale, Whale Two, was a gift that kept on giving. Once it swam past the Australian EEZ border, it kept cruising towards the Sub-Antarctic Front beyond 45°S. By early January 2022, it was clear that Whale Two was not a typical pygmy blue whale who travelled just between Indonesia and Australia. No, it might actually belong to a different population that use the Sub-Antarctic waters as part of their home range. If this hypothesis is true, it brings a lot of responsibility to Indonesia, for we are now responsible not just for the Australian population of pygmy blue whales. Indonesia as a tropical archipelagic nation is also responsible for the Sub-Antarctic population of pygmy blue whales.

On 17 January 2022, Whale Two's satellite transmission stopped. Although the transmission only lasted for two months with just less than 6,000 kilometres travelled, Whale Two's trajectory was intriguing enough that I thought a further investigation was warranted.

Edy Setyawan, one of my fellow researchers, analysed Whale Two's tracking data using a method called movement persistence index to understand what the whale did when the satellite transmission stopped. Had it reached its final summer ground destination and was just hanging around those waters by then? No, apparently Whale Two was still travelling when it stopped transmitting its journey; it had not reached its final destination yet. It also means that it's likely that this whale might have travelled further southwest (or west) towards the waters east of Kerguelen Island and Heard and McDonald Islands. In short, its travel itinerary might have included cruising the waters between the Sub-Antarctic Front and the Polar Front, an ocean front region that has not been associated with pygmy blue whales.

With Whale Two's transmission loss, our attention then turned to Whale One. In mid-January 2022, this whale was still in the waters between the Subtropical Front and the Sub-Antarctic Front south of Western Australia. Yet, as the months progressed, it travelled further south, enjoying the ecological luxury that the Sub-Antarctic Front has to offer. In April 2022, Whale One even reached 49°S, officially making it the first recorded pygmy blue whale to ever make it that far. Afterwards, it lingered around the waters between 47°S and 49°S before deciding to travel back north, back to the western waters of Western Australia. The satellite tracking battery lasted long enough to show how this whale kept cruising northwards along Western Australia and beyond. On 2 June 2022, after almost two hundred days of travel and more than 13,600 kilometres travelled, Whale One neatly ended its transmission just outside Indonesia's EEZ, presumably as it re-entered the Indonesian waters. Not only was Whale One the first pygmy blue whale from Indonesia to enter the Sub-Antarctic waters; it also was the first observed pygmy blue whale to conduct almost a full round of migration between Savu Sea, Western Australia, and the Sub-Antarctic waters.

From national governance to personal spiritualism: Rethinking our relationship with whales

Satellite telemetric research on the pygmy blue whale population that use the eastern Indonesian waters and western and southern Australian waters has been produced numerous times by Australian-based researchers. Since 2009, satellite tags deployed to 37 pygmy blue whales in the southern and western Australian waters revealed that these whales travelled from Australia to Banda Sea via Savu Sea and its adjacent waters (Double et al., 2014; Möller et al., 2020; Thums et al., 2022). One of the whales even conducted a return journey of more than 15,000 kilometres in 271 days (Möller et al., 2020).

Yet, none of those pygmy blue whales was ever observed to venture beyond Australia's southern EEZ. Whale One and Whale Two of our Savu Sea survey were indeed the first recorded pygmy blue whales to venture out of the Australian waters into the high seas, touching the Sub-Antarctic Front as they swam. Through more than six months of travel, Whale One has shown the importance of the waters between the Subtropical Front and the Sub-Antarctic Front south of Western Australia for its sub-species. Despite only having two months' worth of satellite transmission, Whale Two has shed more than enough light on the importance of the Sub-Antarctic waters as part of the journey of its sub-species. It is also likely that Whale One and Whale Two were part of two different populations that used the Banda Sea as part of their home ranges.

Therefore, if we want to protect this sub-species, Indonesia should go beyond improving the conservation governance of Savu Sea and Banda Sea. Indonesia needs to understand that these whales are not just Indonesian whales, and that Indonesia's management decisions will impact marine life in Australia and the Antarctic as well. Nevertheless, research has shown that the Indonesian government is ill-prepared in protecting this sub-species (Sahri et al., 2022). I sadly concur. Despite efforts in eradicating corruption in the last few decades (e.g., Putri et al., 2024), there remain issues with inter-ministerial jurisdiction conflicts and procedural complications that prevent fast action to protect marine mammals, not to mention ignorance amongst the citizens (Mustika et al., 2022).

Current conflicts in human systems of governance can hinder our efforts to protect nature (Smith et al., 2003; Yu et al., 2010). In the Indonesian context, it is not easy for veterinarians to determine the cause of death of a stranded whale, dolphin, or dugong due to jurisdiction conflicts between the Ministry of Marine Affairs and Fisheries and the Ministry of Environment and Forestry, as well as the complicated procedures in transporting genetic materials between provinces in Indonesia (Mustika et al., 2022). One anecdote exemplifies some of these cultural difficulties. I once witnessed an online conversation about a dead sperm whale stranded on an island in Southeast Sulawesi Indonesia on 19 November 2018 (Whale Stranding Indonesia, 2018, Entry 371). As revealed by a necropsy, the poor whale might have died due to the plethora of marine debris it had accidentally ingested in its efforts to feed itself. Amongst the items found were plastic bottles, many plastic packages of snacks, and a single flip-flop sandal. I remember clearly reading online commentary about the incident; an Indonesian commented that if the sandal was not made in Indonesia, then it was not their fault. Ironically, the sandal was made in Indonesia (the brand was a famous Indonesian brand). Other blue humanities scholars have also written of this sperm whale stranding (Oppermann, 2023, p. 31).

Positive changes are happening. For instance, in Australia, supermarket giants have started to use brown bags and now charge for the use of plastic bags. Many states in Australia, including Queensland, have banned single-use plastics (National Retail Association, 2023). I give credit for this progress, but we still need to do more about the single-use plastics used for vegetables, fruits, and meat in the supermarket, down to small items such as the fish-shaped soy sauce we obtain from sushi shops. When not handled properly, these single-use plastic items could easily end up in the ocean, straight into the stomachs of marine animals (including but not limited to whales, of course), giving them painful deaths in the long run (Eisfeld-Pierantonio et al., 2022). On the commercial side, Australia also has improvements to make. Some oil and gas explorations, like the Woodside gas project in Western Australia – smack-bang in the middle of what Greenpeace Australia correctly dubbed the 'whale migration superhighway' (Greenpeace Australia, 2023) – or Schlumberger-SLB and TGS in Victoria and Tasmania (Australian Marine Conservation Society, 2023), are disturbing 'developments' that appear to favour profit over nature.

When I think about Whale One and Whale Two, it is not only as a cetologist. My relationship with these whales is also informed by my spirituality. As a person raised in Eastern philosophies, particularly Hinduism and Balinese traditional knowledge, I generally do not see human needs as above the needs of other non-human species. The Balinese people see, or are supposed to see, humans as part of nature. Rooted in Hinduism, the Balinese have various ceremonies, both for humans and nature. Human-related ceremonies include birth ceremony, turn-of-age ceremony, marriage ceremony, and death ceremony (tourists would recognise the 'Ngaben' or the traditional Balinese cremation ceremony). The Balinese have a whole set of ceremonies for nature, among others Tumpek Uduh (dedicated to the protection of vegetations) (Bali Spirit, 2023b) and Tumpek Kandang (dedicated to the protection of animals) (Bali Spirit, 2023a). The sacred status of

Balinese temples provides an unexpected traditional conservation framework for local wildlife, including birds, monkeys, and other animals that live around the temples. One such temple related to marine life is the Uluwatu Temple in south Bali. Dedicated to Lord Shiva in His role as Pasupati (Lord of the animals), this temple oversees the southern waters of Bali, abundant in whales and dolphins (Mustika et al., 2021).

Another religious event that serves both the human and natural dimensions is 'Nyepi' or the Silent Day. The idea of Nyepi was to welcome the New Year in silence and contemplation, thus the importance of silent meditation on that day. To this end, every Balinese is encouraged not to work, travel, eat, or speak for 24 hours. The absence of travels during Nyepi Day extends to the air and sea where Bali's Ngurah Rai International Airport and seaports are closed for the day. To encourage a deeper state of meditation, the Balinese are banned from turning on lights for 24 hours and they generally comply with the ban (Karang et al., 2021).

Thus, the island comes to a halt on Nyepi Day, where nothing but the sound of nature such as bird songs, crickets, and the wind and waves can be heard. Although it was originally intended for meditative purposes, the absence of activities also entails the absence of travelling and a major reduction of our carbon footprint for that day. Nyepi also significantly reduces ocean noise, which is crucial in creating a stress-free environment for whales and dolphins around Bali (Williams et al., 2018). Ergo, Nyepi Day also serves as a powerful conservation event.

The underlying principles of religious ceremonies in Bali are to honour the relationships between humans and nature. Yet, in an age where profit seems to be the goal of everything we do, instead of being a tool to improve everyone on Earth, it is quite hard to uphold this philosophy without being considered as anti-development, or even as a person insensitive to the needs of other fellow human beings. I understand some of us are trying to adhere to the principles of sustainable development, but in the case of Australia and Indonesia, I have little doubt that we have not been doing enough.

Questions of the whales

Following the transmission loss of Whale Two in January 2022, we also lost the transmission of Whale One in June 2022, so we know nothing about them now. Since my fellow researchers and I are currently preparing a manuscript on the tagging analyses of the two whales, out of necessity I found myself looking at the track map every now and then. However, it is also out of curiosity that I kept marvelling at these tracks.

As a scientist, I often question myself and ask, are they still alright now? If they are females, are they with calves now? Or has my own species destroyed them somehow? I hope they are still alive, not stranded somewhere in Indonesia, with injuries from vessel strikes or with reduced immunities due to plastics ingestion. And if they are still alive, at the risk of humanising a non-human species, I do have these questions. What are they thinking about us? Are we anything more than an irresponsible predator?

I won't know the answers to any of those questions. But I certainly will try to protect their kin from my own species' folly.

Acknowledgements

I would not have been able to write this chapter without the support of Reef Check Indonesia (where I became the marine megafauna consultant during the course of this project) and the Kupang National Marine Protected Area Authority (Balai Kawasan Konservasi Perairan Nasional – BKKPN – Kupang). My sincere gratitude to the following colleagues who provided all support for the project: I. Made Jaya Ratha, Rusydi, Muhammad Offal Prinanda, Edy Setyawan, and the amazing crew of the *Pindito* (particularly but not limited to Edi Frommenwiler and Oom Apé). This project was funded by ICCTF (the Indonesian Climate Change Trust Fund) Bappenas for the COREMAP-CTI World Bank programme with the contract number ICCTF: 003/SPK/PPK-06-14/PH-ICCTF/08/2020.

Note

1 For more on Reef Check Indonesia, see https://reefcheck.or.id/

References

Australian Marine Conservation Society. (2023). *Monster seismic plastic plans in southern Australia's waters threaten whales & marine life*. Australian Marine Conservation Society. Retrieved 25 August 2023 from https://www.marineconservation.org.au/monster-seismic-blasting-plans-in-southern-australias-waters-threaten-whales-marine-life/

Bali Spirit. (2023a). *TUMPEK KANDANG – The Celebration of Animals*. Bali Spirit. Retrieved 8 August 2023 from https://www.balispirit.com/community/ceremony-family/tumpek-kandang

Bali Spirit. (2023b). *Tumpek Uduh Ceremony*. Bali Spirit. Retrieved 8 August 2023 from https://www.balispirit.com/community/ceremony-family/tumpek-uduh-ceremony

Branch, T. A., Stafford, K. M., Palacios, D. M., Allison, C., Bannister, J. L., Burton, C. L. K., . . . & Warneke, R. M. (2007). Past and present distribution, densities, and movements of blue whales *Balaenoptera musculus* in the Southern Hemisphere and northern Indian Ocean. *Mammal Review*, *37*(2), 116–175.

Cassoff, R. M., Moore, K. M., McLellan, W. A., Barco, S. G., Rotstein, D. S., & Moore, M. J. (2011). Lethal entanglement in baleen whales. *Diseases of Aquatic Organisms*, *96*, 175–185.

Double, M. C., Andrews-Goff, V., Jenner, K. C. S., Jenner, M.-N., Laverick, S. M., Branch, T. A., & Gales, N. J. (2014). Migratory movements of pygmy blue whales (*Balaenoptera musculus brevicauda*) between Australia and Indonesia as revealed by satellite telemetry. *PLOS One*, *9*(4), e93578.

Eisfeld-Pierantonio, S. M., Pierantonio, N., & Simmonds, M. P. (2022). The impact of marine debris on cetaceans with consideration of plastics generated by the COVID-19 pandemic. *Environmental Pollution*, *300*, 118967.

Greenpeace Australia. (2023). *Woodside's seismic blasting plans threaten to disrupt whale migration superhighway*. Greenpeace Australia. Retrieved 14 July 2023 from https://www.greenpeace.org.au/news/woodsides-seismic-blasting-plans-threaten-to-disrupt-whale-migration-superhighway/

Karang, I. W. G. A., Ceria, A., & Lynham, J. (2021). Detecting religion from space: Nyepi Day in Bali. *Remote Sensing Applications: Society and Environment*, *24*, 100608.

Leslie, M. S., Perkins-Taylor, C. M., Durban, J. W., Moore, M. J., Miller, C. A., Chanarat, P., . . . & Apprill, A. (2020). Body size data collected non-invasively from drone images indicate a morphologically distinct Chilean blue whale (*Balaenoptera musculus*) taxon. *Endangered Species Research*, *43*, 291–304.

Mustika, P. L. K., Birtles, A., Everingham, Y., & Marsh, H. (2013). The human dimensions of wildlife tourism in a developing country: Watching spinner dolphins at Lovina, Bali, Indonesia. *Journal of Sustainable Tourism*, *21*(2), 229–251. https://doi.org/10.1080/09669582.2012.692881

Mustika, P. L. K., Birtles, A., Everingham, Y., & Marsh, H. (2015). Evaluating the potential disturbance from dolphin watching in Lovina, north Bali, Indonesia. *Marine Mammal Science*, *31*(2), 808–817.

Mustika, P. L. K., High, K. K., Putra, M. I. H., Sahri, A., Ratha, I. M. J., Prinanda, M. O., . . . & Kreb, D. (2022). When and where did they strand? The spatio-temporal hotspot patterns of cetacean stranding events in Indonesia. *Oceans*, *3*(4), 509–526.

Mustika, P. L. K., Williams, R., Kadarisman, H. P., Purba, A. O., Maharta, I., Rahmadani, D., . . . & Dewantama, I. (2021). A rapid assessment of the marine megafauna biodiversity around South Bali, Indonesia. *Frontiers in Marine Science*, *8*.

Mōller, L. M., Attard, C. R., Bilgmann, K., Andrews-Goff, V., Jonsen, I., Paton, D., & Double, M. C. (2020). Movements and behaviour of blue whales satellite tagged in an Australian upwelling system. *Scientific Reports*, *10*(1), 1–19.

National Retail Association. (2023). *Single-use plastic items ban*. National Retail Association. Retrieved 25 August 2023 from https://qldplasticsban.com.au

Oppermann, S. (2023). *Blue Humanities: Storied Waterscapes in the Anthropocene*. Cambridge University Press.

Putri, C. M., Argilés-Bosch, J. M., & Ravenda, D. (2024). Creating good village governance: An effort to prevent village corruption in Indonesia. *Journal of Financial Crime*, *31*(2), 455–468. https://doi.org/10.1108/JFC-11-2022-0266

Reeves, R. R., McClellan, K., & Werner, T. B. (2013). Marine mammal bycatch in gillnet and other entangling net fisheries, 1990 to 2011. *Endangered Species Research*, *20*(1), 71–97. https://doi.org/10.3354/esr00481

Sahri, A., Jak, C., Putra, M. I. H., Murk, A. J., Andrews-Goff, V., Double, M. C., & Van Lammeren, R. J. (2022). Telemetry-based home range and habitat modelling reveals that the majority of areas important for pygmy blue whales are currently unprotected. *Biological Conservation*, *272*, 109594.

Smith, R. J., Muir, R. D. J., Walpole, M. J., Balmford, A., & Leader-Williams, N. (2003). Governance and the loss of biodiversity. *Nature*, *426*(6962), 67–70.

Thums, M., Ferreira, L. C., Jenner, C., Jenner, M., Harris, D., Davenport, A., . . . & Attard, C. R. (2022). Pygmy blue whale movement, distribution and important areas in the Eastern Indian Ocean. *Global Ecology and Conservation*, *35*, e02054.

Whale Stranding Indonesia. (2018). *Whale Stranding Indonesia*. Whale Stranding Indonesia. Retrieved 15 February 2016 from http://www.whalestrandingindonesia.com/new-database.html

Williams, R., Erbe, C., Dewantama, I. M. I., & Hendrawan, I. G. (2018). Effect on ocean noise: Nyepi, a Balinese day of silence. *Oceanography*, *31*(2), 16–18.

Yu, D. W., Levi, T., & Shepard, G. H. (2010). Conservation in low-governance environments. *Biotropica*, *42*(5), 569–571.

13

HMS *PANDORA* AND THE SEA

Tracing eighteenth-century Polynesian artefacts and their entanglement with the Pacific Ocean

Jasmin I. Günther

> She still continued to gain upon us so much that under an hour and a half after she had struck there was eight feet of water in the hold, and we perceived that the ship had beat over the reef where we had 10 fathoms water. We let go the small bower and veered away the cable and let go the best bower under foot in 15 fathoms water to steady the ship. At this time the water only gained upon us in a small degree and we flattered ourselves for some time that by the assistance of a top sail which we were preparing and intended to haul under the ship's bottom we might be able to free her of water, but these flattering hopes did not continue long, for as she settled in the water the leaks increased and in so great a degree that there was reason to apprehend that she would sink before daylight.
>
> Pandora*'s captain, Edward Edwards,*
> *in his report on the events of 28 August 1791*

In late August 1791, the British vessel HMS *Pandora* reached the outer Great Barrier Reef off the Queensland coast in Australia – dangerous and unknown waters for her crew, which hoped to find a safe passage and return home via the Torres Strait. The seamen, under the command of Captain Edward Edwards, had left England nine months prior and had traversed the Pacific Ocean on a single mission: to search for the mutineers of the *Bounty*, whom the Admiralty wanted to see captured and brought back to England for trial and punishment.

The beginning of the crew's mission had been promising, as 14 of the 25 mutineers were arrested soon after *Pandora*'s arrival at the island of Tahiti on 23 March 1791 (Gesner, 2000, p. 5). However, when no specific information on the location of the remaining men or the *Bounty* could be gathered, the ship made sail and continued on a three-month search of the major Polynesian Islands west of Tahiti and the Society Islands, including the Cook Islands, Samoa, and Tonga, before it was

DOI: 10.4324/9781003365501-18

FIGURE 13.1 'HMS *Pandora* in the act of foundering'. Etching by Robert Batty (1789–1848), from an original sketch by Peter Heywood (1772–1831). Originally published in *The Eventful History of the Mutiny and Piratical Seizure of HMS Bounty: Its Causes and Consequences by Sir John Barrow, 1831*, London: John Murray.

Source: Public domain; State Library of Queensland.

decided to set a course home. Yet, *Pandora* would never make it back to England: on 28 August, the vessel ran aground on a submerged reef as the crew attempted to navigate through the Great Barrier Reef. Within an hour and a half, there were over 2 metres (8 feet) of water in the hold, leading to frantic efforts by the seamen to avert further damage. The crew battled throughout the night, but it eventually became clear that *Pandora* could not be saved, and Edwards gave the order to abandon ship (Gesner, 2016, p. 37). While the survivors escaped to a nearby sand cay and prepared themselves for a long and difficult journey home, *Pandora* sank to the bottom of the ocean and would rest there for 186 years.

When the wreck was rediscovered in November 1977, the retrieval of over six thousand objects from the site over the course of two decades and nine seasons of excavation was set in motion. These objects found their permanent home at the Museum of Tropical Queensland (now Queensland Museum Tropics)[1] in Townsville, Australia, where they are stored and partly presented today. Among them are parts of the ship, equipment, tools, instruments, personal belongings of the crew members, natural history specimens, as well as artefacts classified as Polynesian material culture, which proved that *Pandora*'s crew had engaged in collection activities (Campbell & Gesner, 2000). Especially after James Cook's explorations, a flourishing market for these so-called 'artificial curiosities' had emerged in Europe and collecting became a well-established activity for everyone with the opportunity to voyage, as even ordinary seamen must have been aware of their value at home (Campbell, 1997, p. 6). Another motivation for collecting might have been the procurement of souvenirs as visible proof of the extensive travel to remote places. In any case, the objects were apparently considered worth acquiring and keeping, even though the additional cargo occupied space, which was always at a premium on a ship.

Previous research and comparative analyses (e.g., Campbell, 1997; Campbell & Gesner, 2000; Gesner, 2000, 2016) have produced valuable insights about the Polynesian artefacts recovered from *Pandora*'s wreck, although mainly from an archaeological perspective with a strong focus on conservation science and reconstructing the past. My research seeks to engage with the collection differently by drawing on anthropological theories and methods. A long-term research stay in Tahiti from March 2017 to February 2018 and two subsequent, shorter visits facilitated the exchange of knowledge about the artefacts, which needed to be reconnected to people and institutions in Polynesia.[2] My aim was not only to better understand the objects in the context of eighteenth-century encounters and exchange, but to explore what relationships they could be part of in the present and future. As traces of the past that outlived those who once made or collected them, the objects continue to have the potential to make connections and generate new stories.

Researching the *Pandora* collection posed particular challenges because many of the objects are partly degraded or incomplete; after 186 years underwater, only certain materials have survived. The Polynesian artefacts recovered from the wreck are composed of stone, shell and mother of pearl, bone, teeth, coconut husk, and wood – materials which were differently affected by the saltwater depending on their physical properties. Indeed, the ocean played a significant role for the objects and humans involved in this story, which may not thus far have been acknowledged, as it is the latter who are usually placed at the centre of people's interest and enquiry. In this essay, I revisit the *Pandora* collection and look at it through a blue lens, examining how water has shaped the artefacts' trajectories and ability to move – and, in consequence, their connections with people.

Above the sea

By the time of *Pandora*'s arrival at Tahiti, the island's inhabitants had seen many foreign ships come and go, as they had traversed the region frequently since the 1760s. The European visitors were generally guided by two theoretical assumptions: the existence of the Northwest Passage, which would enable travels around or through the North American continent, and that of the southern continent Terra Australis, which was thought to act as a counterbalance to the land masses of the northern hemisphere (Raj, 2000, p. 83). From their perspective, it might have been tempting to see the inhabitants of the Pacific islands as passive and static recipients of these visits. Yet, Oceanic people used the ocean for their own purposes and maintained various relationships with each other over great distances. In fact, their story of movement and dwelling in the region is a long one, stretching back thousands of years (Hau'ofa, 1993, pp. 7–9; Lansdown, 2006, pp. 1–4).

The Pacific Ocean had been opened up to European navigation during the sixteenth century, but many of its islands could not be precisely located again during subsequent voyages, as the procedures for taking bearing measurements lacked standardisation (Raj, 2000, p. 83). The sea was 'as much somewhere to be endured as somewhere to be explored' (Mack, 2011, p. 16) – a seemingly endless space characterised by the unknown, in which explorers could encounter both great discovery and danger. Ships enabled people to move across this space and connect to faraway places, but their success at conquering the oceans was only partial: at times, ships and lives were lost and claimed by the sea, as was the case for HMS *Pandora*.

In their introduction to *Material Ecocriticism*, editors Serenella Iovino and Serpil Oppermann discuss the entanglements of people with more-than-human forces, stating that 'it is quite arduous for humans to declare their agentic independence in a hybrid, vibrant, and living world' (2014, p. 3). When looking at *Pandora*'s voyage and her wreckage in particular, the tangles of cultures and natures become very apparent: despite the crew's best efforts to save the ship, *Pandora*'s line of travel abruptly came to a halt on the outer Great Barrier Reef. Now taking a vertical direction, she sank to the bottom of the Pacific Ocean and became a dot on the map, reconfiguring the physical movements across space of both the ship and her crew. In consequence, the *Pandora* artefacts parted ways with their kind, as their means of transport did not reach its intended destination. Other ships made their way back to England and caused the circulation of similar Oceanic 'curiosities', which were subsequently donated, sold, auctioned, and swapped. Since collecting and museums are closely connected to each other, many of them were presented to such institutions and entered their collections; indeed, eighteenth-century explorations of Oceania set in motion the wide-spread collection and displacement of objects. As the decades passed, the objects experienced phases of neglect and (revived) interest that were accompanied by various attributions and categorisations. Interestingly, the Polynesian artefacts from HMS *Pandora* – although separated from these developments and histories – eventually took a similar path as they, too, found their way into a museum, albeit in Australia instead of Europe.

The ocean has come to the fore as a major participant in *Pandora*'s story, in which human agency – to borrow Iovino and Oppermann's words – is only part of the picture. Of course, roughly two centuries later, humans would return to the site and pull thousands of objects out of the water to, again with great effort, salvage and preserve what had survived. I will return to this important point after a closer look at what occurred below the surface, while *Pandora* remained underwater.

Below the surface

The wreckage of HMS *Pandora* and the significant time that elapsed until her rediscovery may be perceived as a period of stagnation, as the ship never returned to England and remained beyond human reach for a very long time. Yet, these events did not lead to a complete cessation of movement: the rather unusual environment had a significant effect on the wreck and its contents. Over time, the exposed upper levels of the vessel began to disintegrate and gradually collapse due to the effects of marine borer activity and water motion, causing heavy durable objects to sink vertically into the sand below (Gesner, 2016, p. 6). Lighter items that were more receptive to water movement likely tumbled onto the seafloor to either side of the hull remains, where they were later buried by accumulating sediments. Other objects may have been swept further away by currents and waves or were trapped under the hull before the stern collapsed on top of them (Gesner, 2016).

Jonathan Rankin and Francis Collins speak of the ocean as an engine of decay in their discussion of cruise ships, 'stripping ships of paint, rusting their bodies and eventually disassembling them completely' (2017, p. 239). Unlike today's ships, which are largely made from steel, eighteenth-century vessels like *Pandora* were predominantly built from wood – a material that is also, although differently, affected by long-term submersion in saltwater and the ocean's constant wave motion. After long periods underwater, bacterial action causes the degradation of cell wall components in all wood, as water-soluble substances (such as starch, sugar, and mineral salts) are leached from the waterlogged material. Over time, cellulose in the cell walls disintegrates, leaving only a lignin to support the wood, which will eventually start to break down as well. As a result, spaces between the cells increase and the wood becomes more porous (Hamilton, 1999). Interestingly, most objects will retain their shape as long as they are submerged in water: once a critical point is passed, the material reaches an equilibrium and will cease this rapid deterioration. Additional years in this environment are therefore considered less threatening than the process of pulling the objects out of the water (personal communication with Sue Valis, Conservator, Museum of Tropical Queensland, 2017).

For the wooden artefacts retrieved from *Pandora*'s wreck, it was therefore the moment of being removed from the ocean after such a long period of time that was crucial and potentially dangerous. If exposed to air untreated, the surface tension created by drying and evaporation of excess water causes the weakened cell walls of waterlogged wood to collapse, resulting in irreversible damage, such as shrinkage, distortion, and cracking (Hamilton, 1999). This kind of damage would arguably make any artefact a less valuable museum object, and great care was applied during the excavations, which occurred during the 1980s and 1990s. Indeed, for the *Pandora* artefacts, conservation commenced on the seabed and continued onboard the expedition support vessel. All objects were kept moist or fully immersed in seawater until they could be transferred to a vehicle to be transported to the conservation laboratory (Gesner, 2016, pp. 124, 147–148).

Other materials were differently shaped by the time underwater, depending on their physical properties and how these react to saltwater and the marine environment. For example, many of the Polynesian stone artefacts in the *Pandora* collection – mainly pounders and adze blades – were covered in marine deposits upon their retrieval from the wreck. I related these encrustations to a specific category of traces, discussed by anthropologist Tim Ingold in his book *Lines: A Brief History* (2007). According to Ingold, traces can, in most cases, be distinguished by being either reductive or additive, defining them as 'any enduring mark left *in* or *on* a solid surface by a continuous movement' (Ingold, 2007, p. 43; emphasis added). To me, the degraded cell wall components of the wood represented reductive traces and the layers of deposits on the stone additive ones. The latter were not removed in the field but treated after safe transportation of the objects to the laboratory, where most of the stone artefacts were mechanically cleaned. Untangling the objects from the water required counteracting its additive and reductive traces, which was a laborious and, at times, slow process.

Out of the water

Once exhumed from the water, the process of preservation begins in earnest. Artefact preservation is – next to collecting, researching, and exhibiting – one of the key areas of responsibility for a museum institution and mainly comprises the field of conservation: the process of documentation, analysis, cleaning, and stabilisation of an object in order to protect it. Because most artefacts would perish without conservation treatment, it is considered essential for their safe study, interpretation, and exhibition as well as the preservation of historic data for present and future generations (Hamilton, 1999). For the materials recovered from *Pandora*'s wreck, it was an important part of their transformation into museum objects.

Different types of materials (organic and inorganic) require different conservation techniques, and a shipwreck collection comprises an additional layer of complexity. Wooden objects that were submerged in saltwater for considerable time are especially challenging, as they often appear well-preserved but are of a very fragile nature due to the above-mentioned degradation of their cell wall components. To stabilise them, they are usually subjected to an elaborate treatment: they are washed and brushed under running deionised water and then placed in a vat containing a solution of synthetic polymer called Polyethylene Glycol (PEG) and solvent (such as water or alcohol), which slowly permeates the wood and replaces the bound water by osmotic diffusion. This process can take several weeks, months, or even years, depending on the object and its condition, size, and species of the wood. Afterwards, the objects are frozen and vacuum freeze-dried – a procedure that involves the conversion of water from its frozen state into a gaseous phase which helps avoid drying stresses and damage. Once dry, any solidified excess PEG visible on the surface is mechanically cleaned (Hamilton, 1999).

Among the Polynesian artefacts within the *Pandora* collection were nine clubs made from wood, some of which were visibly affected by their time underwater and partly degraded, while others appeared well-preserved and in remarkable condition in light of the artefacts' history. Through human intervention, the clubs' internal structure was completely changed, and the artefacts were reinforced and stabilised with Polyethylene Glycol that slowly permeated the wood and replaced the bound water. Because of this treatment and the permanent curation in climate and temperature-controlled environments, their 'life expectancy' has been prolonged – a common trait among museum objects, which are expected to be preserved and therefore bound to strict regulations and conservation processes.

Because of their continued existence, museum objects can function as material embodiments and representations of a specific place, time, people, or culture (Knowles, 2013, p. 229), including those collected in eighteenth-century Oceania.[3] Artefacts made from stone, in particular, have become popular objects of research, as they can be extremely durable and potentially accompany humans for long periods of time. For this reason, they are often valued for their abilities to tell us something about the past, such as changes over time and movements across space. Polynesian stone adzes, for instance, have been the subject of scholarly writings for over one hundred years and inspired the creation of island-group-specific typologies based on their formal characteristics, which were generally based on the artefacts available in museums. Although these typologies are today viewed more critically, they were still helpful in researching the *Pandora* collection: drawing on comparable artefacts and typologies (Figueroa & Sanchez, 1965), the majority of its 26 stone adze blades could be traced back to the Society Islands and, more specifically, the island of Tahiti, where *Pandora*'s crew spent several weeks. This hypothesis was supported by experts in French Polynesia, including at the Musée de Tahiti et des Îles (Te Fare Iamanaha), whose collections comprise a large number of both stone adzes (*to'i*) and pounders (*penu*) (Lavondès, 1976). The five pounders recovered from the wreck were also identified to be from the Society Islands.

Arguably, any identification and comparative analyses were facilitated by an unobscured view of the *Pandora* objects due to the removal of the encrustations that concealed the stone artefacts. Further conservation treatment of an adze blade – according to the condition reports available – usually consisted of mechanically removing deposits and cleaning the object, followed by stages of washing, rinsing, and drying. Once freed from the marine deposits, the adze blades resembled their 'original state' more closely. However, the artefacts were not necessarily 'complete', since they were recovered as individual, loose items and not as hafted adzes. While it is possible that the adzes were collected (or stored) as such, it is likely that the blades and wooden handles were separated from one another because the plant fibre cords that were once holding them together had dissolved during their time underwater. These disintegrated materials draw attention to both the destructive forces of the ocean and the absences in the collection which, in many instances, generated a sense of wonder, speculation, and investigation as the remnants were and are pieced together.

As described earlier in the chapter, the wooden clubs from HMS *Pandora* were likewise affected by their long-term submersion in saltwater. Still, most of them could be compared to similar eighteenth-century artefacts from Tonga with the help of various typologies (Mills, 2009), based on their shape and different kinds of decorative design elements. The Polynesian artefacts within the *Pandora* collection differ from their 'pristine' counterparts as a result of their entanglement with the Pacific Ocean and yet, this very entanglement can be considered a critical part of the narrative of these objects and the telling of new stories with these material presences and traces.

Across time

Identified and allocated to categories and object groups, the *Pandora* artefacts were put in place, which rendered them immobile in some respect, as museum records rarely change. In addition, once settled into a storage or exhibition space, physical movements out of the museum, and even within it, hardly ever occur; the objects' ability to move across time *and* space has been significantly altered. In their removal from the ocean, they are reconceived and reimagined according to human epistemologies.

Anthropologist Amiria Henare argues that the mobility of museum objects across space is 'restricted precisely *in order to enhance their ability to move through time*' (2005, p. 9; emphasis in the original). Because they are kept to preserve something of the past, the museums' responsibility and care for them includes conservation treatment. As demonstrated earlier in the chapter, the latter ensures their continued existence and thereby enables them to generate and perpetuate connections with people. Unlike an understanding of museum objects as things 'removed from other times and other places, [which] are neatly labelled, catalogued and packed away out of sight, rarely displayed and infrequently studied' (Byrne et al., 2011, p. 4), Henare holds a more optimistic view and emphasises their potential for present and future generations.

For the Polynesian artefacts from HMS *Pandora*, new connections to Polynesia were possible *because* materials have survived and were transformed into museum objects; however, with their movements across space restricted, these connections are now largely made in digital form. The conversations with my research partners in Tahiti often involved an engagement with the artefacts through photographs and a digital catalogue that I had compiled prior to my arrival in French Polynesia. Yet, over the course of my research, I also had to acknowledge the apparent absences associated with the collection: despite the recovery and preservation of these material presences, many other materials and knowledge once attached to the objects were lost with the sinking of the ship and its extended period of time underwater.

The first-hand accounts of *Pandora*'s voyage available to us today unfortunately give little information about the crew's encounters and exchanges with the inhabitants of the islands they visited; in consequence, the makers and previous owners of the Polynesian artefacts remain unidentified, and the exact reasons for their collection a mystery. Potential collectors among *Pandora*'s crew can only be speculated about despite compelling theories based on the location of the excavated artefacts within the wreck, which were correlated to the construction plans of the ship (Gesner 2000, 2016). In the end, every approach has its boundaries, and, at a certain point, we run into what James Flexner called a great limitation of archaeology and many other historical sciences: the problem of equifinality, meaning 'the situation in which multiple historical explanations are equally plausible for observed phenomena in the present' (2016, p. 176). The entanglement of *Pandora* with the ocean complicated people's ability to learn more about the artefacts and reconstruct the past.

Given these limitations, I focus more strongly on the present significance and future potential of the *Pandora* collection. My stay in Tahiti allowed me to follow comparable objects created today, which indicated that Polynesian artefacts from the eighteenth century – and especially their shapes and forms – play a role in the making of contemporary (art)works and continue to move and inspire people. This realisation led to an exploration of the ways in which people in Oceania today relate to their dispersed cultural heritage in museums worldwide. Stone pounders, for instance, are visible in Tahitian art and in everyday life, including in galleries, at markets, and in people's homes. Some of these objects were made to resemble their eighteenth-century counterparts, while others have taken other forms, were crafted from different materials, and were made for different purposes. Miriama Bono, the director of the Musée de Tahiti et des Îles at the time, named pounders as an example of an object that has become emblematic of Polynesian culture during one of our conversations about the *Pandora* collection (personal communication, 2018).

These objects and their makers set in motion a shift of my focus: while the early stages of research were strongly influenced by the concept of mapping movements and the hope of finding stories directly related to the *Pandora* artefacts, more time and space were gradually given to acts of making. Contemporary Oceanic art is diverse, innovative, and constantly changing, and its creation is tied to a thriving and vibrant industry. While *Pandora* rested underwater for 186 years, the making of objects and art in the region never ceased – shapes and forms moved on with people, and stories continued to be told.

As part of this view and a responsibility felt towards the HMS *Pandora* collection, I developed the idea of an exhibition project in collaboration with Tahiti-based artists, who created new artworks in direct connection to the Polynesian artefacts recovered from the wreck.[4] The hope was to generate a more tangible link between the artefacts, now in Australia, and people in Oceania. Coincidentally and independently of each other, artists Jean-Daniel Tokainiua Devatine and Hiro Ou Wen made the decision to focus on the fishing tackle within the collection and created artworks from mother of pearl, which led to a strong focus on ocean-related objects and themes in the exhibition.[5] With the exhibition, I explored the creation and telling of new stories as alternative ways of engaging with museum objects, their histories, and their relevance.

In *Being Alive: Essays on Movement, Knowledge, and Description* (2011), Ingold made an interesting distinction between storytelling and classification, which I could relate to the *Pandora* artefacts, now kept at a museum. According to Ingold, in classification, elements are slotted into place based on intrinsic characteristics that are given quite independently of the context in which they are encountered and of their relations with the things that preceded, presently surround, and follow them. In a story, by contrast, it is precisely through this context and these relations that elements are identified and positioned, and stories thereby draw together what classification splits apart (Ingold, 2011, p. 160). In the storied world, things do not exist but occur and bind up in another's story upon meeting, with every such binding being a place or a topic.[6] Ingold concludes that, in this world of movement and becoming, we can 'understand the nature of things only by attending to their relations, or in other words, by telling their stories' (2011, p. 160).

Telling stories with the Polynesian artefacts from HMS *Pandora* involved an acknowledgement of how deeply the collection was shaped by the ship's wreckage and subsequent time at the bottom of the Pacific Ocean. A striking example for this is adze blade MA7638, which is covered in encrustations (see Figure 13.2). While most other adze blades were mechanically cleaned during their conservation

FIGURE 13.2 Polynesian stone adze blade (MA7638).

Source: Photograph by the author. Image courtesy of the Queensland Museum Network.

treatment, the marine deposits were not removed from this object – perhaps to demonstrate how the artefacts were affected by the underwater environment. Upon discussing images of MA7638 with my research partners in Tahiti, I enquired if the encrustations should have been removed, and the response was always 'No'. Instead, it was accepted that the current state of the artefact was simply a result of its history and part of its story, in which the ocean played a significant role and left its traces. With the stone artefacts, now partially covered in deposits, and the wooden clubs, permeated with PEG, shapes and patterns still move on. Preserved, like other eighteenth-century Polynesian objects in museums today, the *Pandora* artefacts – as well as their entanglement with the ocean, histories, and contemporary conservation – enable connections and new relationships across the sea.

Notes

1 The name change occurred in 2024.

2 My PhD project was funded by James Cook University, Australia, and Aarhus University, Denmark. The research and resulting exhibition were generously supported by the Queensland Museum Network and staff at the Museum of Tropical Queensland (now Queensland Museum Tropics), the Musée de Tahiti et des Îles (Te Fare Iamanaha), the Centre des Métiers d'Art de la Polynésie française, and the contributing artists. I thank them, my supervisors, and the many people in French Polynesia and beyond, who have given so much of their time and kindly shared their advice, knowledge, and art.

3 At the same time, the objects should not be understood as manifestations of an idealised, unchanging 'Pacific Island material culture' but rather as potential evidence of highly dynamic processes that already existed prior to the Europeans' arrival in the region (Hetherington & Morphy, 2009, p. 8).

4 The exhibition *Making Connections – French Polynesia and the HMS Pandora Collection* was on display at the Museum of Tropical Queensland from August to December 2019.

5 Tokainiua Devatine and I further elaborated on the collaborative exhibition project and his artwork in our article 'Making connections and building bridges' (2021).

6 The concept of binding is complex; for more, please see Ingold (2011). In addition, the concept has been discussed in relation to Oceanic art by various authors, including Babadzan (2003) and Küchler (1999). In my thesis (Günther, 2021), I connect the concept to the *Pandora* collection in the chapter on plant fibres.

References

Babadzan, A. (2003). The Gods stripped bare. In C. Colchester (Ed.), *Clothing the Pacific* (pp. 25–50). Oxford, New York: Berg.

Byrne, S., Clarke, A., Harrison, R., & Torrence, R. (Eds.). (2011). *Unpacking the Collection. Networks of Material and Social Agency in the Museum*. Springer.

Campbell, J. (1997). Eighteenth-century wooden clubs from HMS *Pandora*: A preliminary analysis. *Bulletin of the Australian Institute for Maritime Archaeology*, *21*(1&2), 1–8.

Campbell, J., & Gesner, P. (2000). Illustrated catalogue of artefacts from the HMS *Pandora* wrecksite excavations 1977–1995. *Memoirs of the Queensland Museum | Culture*, *2*(1), 53–159.

Devatine, J.-D. T., & Günther, J. (2021). Making connections and building bridges. *Sūdō Journal*, *3* ('Burning bridges'), 107–117. ResearchOnline@JCU. https://researchonline.jcu.edu.au/69674/

Figueroa, G., & Sanchez, E. (1965). Adzes from certain islands of eastern Polynesia. In T. Heyerdahl & E. N. Ferdon (Eds.), *Reports of the Norwegian Archaeological Expedition to Easter Island and the East Pacific – Volume 2* (pp. 169–254). Monographs of the School of American Research and the Kon-Tiki Museum.

Flexner, J. L. (2016). Archaeology and ethnographic collections: Disentangling provenance, provenience, and context in Vanuatu assemblages. *Museum Worlds 4*(1), 167–180.

Gesner, P. (2000). HMS *Pandora* project – a report on Stage 1: Five seasons of excavation. *Memoirs of the Queensland Museum | Culture, 2*(1), 1–52.

Gesner, P. (2016). Pandora project Stage 2: Four more seasons of excavation at the *Pandora* historic shipwreck. *Memoirs of the Queensland Museum | Culture, 9*.

Günther, J. I. (2021). *The Movement of Things: Tracing Eighteenth-Century Polynesian Artefacts from HMS Pandora* [Doctoral thesis, James Cook University and Aarhus University]. ResearchOnline@JCU. https://researchonline.jcu.edu.au/69658/

Hamilton, D. L. (1999). *Methods of Conserving Archaeological Material from Underwater Sites*. Conservation of Archaeological Resources. Texas A&M University.

Hau'ofa, E. (1993). Our sea of islands. In E. Waddell, V. Naidu, & E. Hau'ofa (Eds.), *A New Oceania: Rediscovering Our Sea of Islands* (pp. 2–18). Suva: School of Social and Economic Development, The University of the South Pacific.

Henare, A. (2005). *Museums, Anthropology and Imperial Exchange*. Cambridge University Press.

Hetherington, M., & Morphy, H. (Eds.). (2009). *Discovering Cook's Collections*. Museum of Australia Press.

Ingold, T. (2007). *Lines: A Brief History*. Routledge.

Ingold, T. (2011). *Being Alive: Essays on Movement, Knowledge and Description*. Routledge.

Iovino, S., & Oppermann S. (2014). *Material Ecocriticism*. Indiana University Press.

Knowles, C. (2013). Artifacts in waiting: Altered agency of museum objects. In R. Harrison, S. Byrne, & A. Clarke (Eds.), *Reassembling the Collection: Ethnographic Museums and Indigenous Agency* (pp. 229–257). School for Advanced Research Press.

Küchler, S. (1999). Binding in the Pacific: Between loops and knots. *Oceania, 69*(3), 145–156.

Lansdown, R. (2006). *Strangers in the South Seas. The Idea of the Pacific in Western Thought*. University of Hawai'i Press.

Lavondès, A. (1976). *La culture matérielle en Polynésie et les collections du Musée de Tahiti et des Îles* [Unpublished doctoral thesis]. Paris Descartes University.

Mack, J. (2011). *The Sea: A Cultural History*. Reaktion Books.

Mills, A. (2009). 'Akau Tau: Contextualising Tongan war-clubs. *The Journal of the Polynesian Society, 118*(1), 7–45.

Raj, K. (2000). 18th-century pacific voyages of discovery, 'big science', and the shaping of an European scientific and technological culture. *History and Technology, 17*(2), 79–98.

Rankin, J. R., & Collins, F. L. (2017). Enclosing difference and disruption: Assemblage, heterotopia and the cruise ship. *Social & Cultural Geography, 18*(2), 224–244.

PART 5
Imagining blue futures

14

COLOURBLINDNESS IN/OF PLACE

Memory, colonial place, and education's ignorance of the blue

Bryan Smith

Venturing out on the ferry across the waters between Townsville and Magnetic Island always puts me in a space of contentment. As a Canadian living in 'Australia', the novelty of tropical life has yet to wear off.

> *I can't believe that this place, one that looks like a vacation spot, is where I now call home.*

The novelty, though, has a travelling companion: a spectre of criticality, one that deservedly haunts my contentment. The ferry, arriving on Yunbenun/Magnetic Island, conjures in me an angry voice:

> *This place is saturated with colonial logics and efforts. This vacation spot is a production of violence.*

My conscience isn't wrong; the island, home to a settler-national park that drapes over the land, is a 'symbol of landed affluence of settler nations', the spatial production of wilderness, and 'a kind of triumph of settler colonial appropriation' (Mar, 2010, p. 76). The park shapes what is possible on the island, and serves to frame, in the popular imaginary, this place as a refuge, a tropical paradise, and a site of unacknowledged dispossession.

Arriving on the island, I am welcomed to 'Magnetic Island', a name of imposition that exists palimpsestically atop an Indigenous toponymy with an opacity that makes it difficult to see through. The island's name is rather uninspired and observational; for Cook, that 'the Compass did not traverse well when near it' (1771/1893, p. 271) warranted the writing into the map a name of 'Magnetic(al)

DOI: 10.4324/9781003365501-20

Island'. Yet, what matters less here is Cook's logic but the layering of an imposed toponym, draped over the space that Cook and the eventual 'Australian' state had and have no legitimate authority over. This island is, then, a troubled place, a site of ontological homelessness (Keys Young, 1998; Watson, 2009) where the state and its figurative progenitors have carefully guarded place-making for invaders, all the while curating a site of preservation to placate its own natural and national (park) needs.

Returning to the mainland, I'm once again reminded of the violence that occurs through everything from the naming (Alderman & Rose-Redwood, 2020) to the concretisation of material and symbolic structures that persistently remind sojourners and residents that this place is annexed space for 'us', not for 'them' (Porter, 2018). These two places (the island and the mainland), taken together, serve as testaments to violence and the artifices of dispossession that encode invasion as benevolent and/or a presumed logical extension of the past into a normalised present.

This consideration of place, though, is partial, failed by conceptual and material lacunae. The journey back to (and from) the mainland didn't simply happen along a plane devoid of meaning; the ocean space between the island and the mainland is imbued with a history, one equally pregnant with meaning, affect, and influence on the place and in the imaginaries of the public. The blue space in between is not just a medium of movement but equally integral to the place of this area. The blue in the middle has stories, a history, and a politics that are equally entangled with the present and how we envision the places of our lives. In this place, those waters between are equal protagonists in the story of here, equally complicit in the regulation and making of white settler place. Yet, in my field of education, the blue is often excluded, with green spaces monopolising critical conceptual and pedagogical work.

This paper wrangles with the aforementioned absence by thinking about three colours – blue, green, and white – and their mutual reconstitution in invader (settler) contexts. As I argue, in my field of education, the green monopolises the conceptual palette in thinking about place, to the almost near exclusion of the blue, a 'colourblindness' that undermines the work of contending white invader supremacy in/of place. This, I argue, is deeply problematic, for the making of white place (and its necessary unmaking) is a blue *and* green affair, with each hued domain an equal 'tint' in the (re)making of normalised invasion. I begin by exploring what place is, particularly as it pertains to the intersections of memory and identity that translate space into a socio-cultural location. From here, I explore how place as a concept in education is notably silent on the blue, concluding with a call for more 'cyan' place thinking in education to properly attend to how place, as a site of white memory and identity, requires critical attention to the mixing of blue and green.

As a point of conceptual clarity before I continue, the use of 'green' here is shorthand for the vast expanse of non-watery spaces, including the brown and red spaces that are extensive in a place like Australia. Equally, 'blue' is shorthand here for watery spaces, those that are increasingly changed by pollution, climate change, and capitalism which shift the colour of watery spaces.

Place, memory, identity: Invasion's unholy trinity

The key theoretical entry point for this paper is the notion of place, particularly how place is the transformation of (blue and green) space into meaningful locales via the impressions of ongoing cultural production and historical inheritance. As Edward Casey (1997) has argued, 'place brings with it the very elements sheared off in the planiformity of site: identity, character, nuance, history' (p. xiii). Put differently, Thomas Gieryn (2000) argues that place is the intersection of location, materiality, and invested meaning or what Anderson (2010) identifies as the assemblage of cultural traces in space. Critically, the congealing of meaning and space into place is not bound by the limits of the green, but rather, pervades all space – blue and green – where history's impress has left its mark.

In contexts of invasion, such as the 'Australian' context from which I write, this process of placial creation was and remains grounded in what Casey (1997) calls 'deplacialization'; the exploration of lands new to invaders was and remains a process inherently about evacuating understandings of place from the epistemological and material landscapes to make way for European ones. Indeed, as Brian Harley has argued, mapping lands (and waters) involved 'engulfing them [Indigenous peoples] with blank spaces' (1992, p. 531), an erasure predicated on false beliefs of lands emptied of civilisation and meaning. This historical and ongoing reconstituting of place to accord with invader geography requires us to consider how place is read and made through the relations of power where forms of oppression intervene to mediate subjective identification, and where individual and structural forces interact (Butler & Sinclair, 2020).

Normalising invader place-making: The role of memory

Turning blue and green spaces into normalised expressions of invader place is made possible, in part, through the use of place as a medium (and product) of collective memory which itself gets encoded into the concrete and imagined places of 'Australia'. As a myopic and instrumental operationalisation of the past, collective memory is a rendition of the past committed to a singular and heroic reading of past experience, one that reflects a dominant group's relationship to its own past and one that shuns ambiguity and difference in favour of nationalising certainty (Wertsch, 2004). Further, this form of memory is relatively static, a condition that allows public memory to serve a critical role in social and collective identity formation (Abel et al., 2018). Through this, collective memory derives an insulation from critique given its political and social value and (re)constitutive function in defining and measuring who 'we' are. Given that it provides a temporally anchored story of 'us', it can be rather impervious to shifts; while collective memory does change (through activism and resistance for instance), it draws on the defensive commitments of those who derive their orientation to the past and self from its hegemonic expression.

In places across 'Australia', collective memory is decidedly textured by invader memorialisation. Drawing on a victomological reading of the past, one of adversity over hostile lands and peoples (Curthoys, 1999), the memory work through and of place is reproductive of a particular representation grounded in nationalist ascendancy, 'civilisation', and prosperity (rooted in an exploitation, natural and corporeal, that is only acknowledged in a cursory sense). This memory can be expressed in/through rather explicit media (such as statues) and more banal features such as the names of our communities (Roediger & Wertsch, 2008; Stanley, 2009). Consider, for instance, street-naming practices, a language of place that maps the past on to an urban space and re-asserts a hegemonic version of the past (Azaryahu, 2009; B. Smith, 2024). Said naming practices write a colonial narrative into the language of the city and the blue places of our lives (consider, for instance, the Tasman Sea that separates Australia from New Zealand, named for explorer Abel Tasman), the result of which is the encoding of a memory into the most mundane features of the landscape (B. Smith, 2018), accomplished in such a way as to appear 'uncontestable' (Rose-Redwood & Alderman, 2011). The blue is a powerful anchor point in this colonial memory work; speaking to the contexts of the naming politics of the Salish Sea, Tucker and Rose-Redwood (2015) point to how the watery spaces serve as a powerful conduit for reinforcing colonial hegemony and state power over all spaces, green and blue.

This spatially encoded memory, and the re-inscription of colonial power, tell the story of the heroes of invasion, those that inaugurated the politics and violence of dispossession that made way for 'settlement'. Heroism is encoded here not innocently but rather, as a heroism of invasion, made normal or (ostensibly) innocent by virtue of alignment with collective colonial memories. It is important to note, however, that invader collective memory is not all-encompassing; while colonial geographies sought to erase Indigenous namescapes, Indigenous toponymies continue to provide a logic and 'sense' to place around the various Indigenous Countries of 'Australia' (Hercus et al., 2002). Indeed, while Indigenous naming practices may be obscured by the opacity of colonial toponymies, Indigenous naming practices continue to circulate as original and lasting names of place, whether in practice or in story.

Finally, the role of place and memory is instrumental to social identity. One's identity, as Casey puts it, is one half of the 'constitutive coingredience' (2001, p. 684) of place. This coingredience gestures towards a complex entanglement of place with those who inhabit and give form to that very place; as argued in this section, this coingredience is mediated by memory, anchoring the relationship between location and self against the backdrop of a memory. Further, the interpolation of subjects within this colonial domain are expressed and measured in relation to taken-for-granted and omnipresent significations about whose (white) place it is and who can or does belong. As Aileen Moreton-Robinson puts it,

> In our quotidian encounters, whether it is on the streets of Otago or Sydney, in the tourist shops in Vancouver or Waipahu, or sitting in a restaurant in New York, we experience ontologically the effects of white possession. These cities signify with every building and every street that this land is now possessed by others; signs of white possession are embedded everywhere in the landscape.
>
> *(2015, p. xiii)*

Those who inhabit this place are asked to take up and imagine their identity in relation to this pervasive signification and are called to recognise some social and cultural identities as more or less congruous with 'our' memory. For these reasons, place is a central political and conceptual focus of decolonising work (Moodie, 2018), important as it is as a key concept in critically contending the 'burial' of Indigenous ontologies by way of invader placial efforts (Watson, 2002).

(Education's) colourblindness

The aforementioned framing of place as an organ for white memory and identity reminds us of the ethical imperative to critically envisage place as guilty in regulating how the spaces of our lives become (violently) meaningful or 'placial'. That said, the conceptual framing of place/identity/memory as a moderating frame for lands may very well be construed as: (a) failing in introducing the blue (as watery spaces can be simply excised from consideration) and; (b) may itself be suggestive of a 're-greening' of placial consideration by focusing on how people derive a sense of self from their green places. Yet, invader geographies, by virtue of their hubristic operation and self-assumed responsibility, become universal and extend across watery and terrestrial space. 'Triumphant settler colonialism is, after all,' Veracini reminds us, 'a violent act against geography: settler colonialism turns someone else's place into space and then into place again' (2017, p. 5). The translation of lands into places stolen by invaders was, and still is, preoccupied with watery and terrestrial spaces.

If we take seriously the critical responsibility to contend what Bonds (2020) calls the 'possessive geographies' of invasion and the making of place in service of hegemonic control, 'place' as a critical and pedagogical concept needs to be used in a way that properly interrogates the complex entanglements of the blue and green. Here, a crucial starting point is to recognise the encompassing idea of Land, what in Indigenous epistemologies has been taken up to be inclusive of blue and green spaces (Sepulveda, 2018; Tuck et al., 2014). Invader geographies separate blue from green, a result of the tendency towards duality that the Western (placial) epistemologies of invasion favour. My own field – education – is similarly shaped by Western imperial ideas and has inherited a penchant for dichotomisation: mind/body, failure/success (pass), learning/not learning. More broadly, though, education has long been a Western affair, drawing as it does on colonial disciplinary knowledge (B. Smith, 2020; Tuhiwai Smith, 1999) to structure its understandings of the body, knowledge, and the world. Just as schools have long taught students how to make and understand a divided world (Willinsky, 1998), it has and remains committed to a divided Earth of aqueous and terrestrial domains, largely favouring the latter as a site of civic, intellectual, and conceptual action.

The inheritance of Western ways of thinking is apparent in place-based educational work that preoccupies itself with the conflation of place with green. For Greenwood, for example, place is a bounded space, where 'its physical scale can range from small indoor corners to sweeping outdoor expanses: mud puddles, bedrooms, backyards, prison cells, back alleys, neighborhoods, farms, watersheds, regions—all of these can be identified and experienced as places' (2013, p. 93). Here, puddles and watersheds exist as the only blue places, both of which are quintessentially terrestrial spaces that hold water and spaces that 'muddy' the lines between blue and green, making the blue worthy of consideration largely as it relates to the green. Elsewhere, in describing how to support the development of a sense of place, Sanger slips into using 'land' as synonymous with place, suggesting that developing a sense of place has, as its first step, 'developing a connection to the land [which] requires experience' (1997, p. 5). Such conceptual confusion is echoed elsewhere, with Stickney asking 'So, what does land- or place-based education look like?' (2020, p. 1076). While Stickney makes reference to blue spaces, the focus on something like a walking focused encounter with place and even the use of the phrase 'pedagogical terrain' (as a metaphorical framing of praxis) gives priority to a green rendering of place as a site of educative potential and possibility.

The work of centring land can sometimes be more subtle, evident through meaningful exclusion of watery spaces. Ho and Chang (2022), for instance, lament the whiteness of conceptions around the outdoors and environment, but do so while preoccupied with the terrestrial spaces upon which whiteness is mapped. Elsewhere, Gruenewald (2008), concerned with similar relations of power and dominance in place, frequently slips into using land in the same breath as place; for example, Gruenewald argues, 'the exercise of power, colonization, imperialism, and exploitation have always involved people *and* land, culture *and* the places that make culture possible' (p. 146, emphasis in original).

In part, some of the concerns here may be a function of ideas of 'place-based' education which is focused on the locales of students and their lives (G. Smith, 2002). Much of this work draws on critical pedagogical or cultural studies scholarship to help students develop more critical readings and relationships with their locales (Israel, 2012; G. Smith, 2002). A critical and cultural focus is an important part of thinking about place, particularly the articulation of memory and identity through place, but a lack of explicit or a common focus on blue spaces risks positioning placial concerns of critical and cultural understanding as they get anchored in the lived spaces of everyday life which, for many, will be largely shaped by green spaces where learners live, work, play, and make knowledge of the world. Indeed, the rather implicit or absent aqueous focus risks discarding blue spaces as influential outlets of meaning (re)making, prompting teachers and students to contend and confront critical, memory, and cultural work in the terrestrial domains of their lives. This results in a critical absence: the importance of water as a constituent and central feature in the makings of place, experience, and memory, a set of connections present in Indigenous ontologies of place (Yazzie & Baldy, 2018). Additionally, there is the risk of seeing and selecting those green spaces that accord well with settler pedagogical practice whereby those places that echo back invasive visions of place – those manicured, organised, and regulated – become places that have a monopoly over the meaning-making process.

It is important to note that the focus on local environments does not preclude blue encounters and to assume this is to, itself, take a 'green stance'. Yet, there is something striking about the notable absence of blue spaces beyond their presence as examples of material objects to explore and not as (re)constitutive of and complicit in place work and the attendant memorial and identity work that is attached and (re)produced through place. The place/green conflation, intended or otherwise, makes for a green pre-eminence, one where blue spaces are ancillary characters in the story we do or can tell and the identities that are (re)produced about/in/of this place. As students begin to see the conflation as normative (or meaningful absence of explicit blue consideration as normative), any critical work to interrogate memory will or could get contained to the material and symbolic artifices of terrestrial spaces. Blue spaces, in effect, risk becoming a part of the absent curriculum, resulting in the exclusion or neglect of watery spaces as central to the memories woven into and of place (Wilkinson, 2014).

Entangling towards cyan: Spectral knowing

I want to turn back to the locale from which I write for a moment. As a migrant, I'm not the only one to journey across watery spaces to this land but, crucially by way of distinction, some of those who have made the journey have not had the luxury of exercising their own volition. In North Queensland, from where I write, invaders' early concerns were initially mediated by a science that understood place as conceptualised, in part, through a lens of tropical medicine. In their early years, invaders' scientific knowledge presupposed corporeal incompatibility; underneath the oppressive sun of the area, widespread concern about the safety of the white body was rampant (Bashford, 2000; Robinson & Wilson, 2008). Deploying a logic of fragility that insulated white bodies from a self-inflicted challenge to their strength, invader discourse simultaneously produced a logic wherein 'they' were incapable of working the lands, but the racialised non-White body was suited, so the thinking went, for the environment. In part, this resulted in the search for bodies to work the plantations of the region, the consequence of which was the slave and indentured labour trade whereby South Sea Islanders were captured and brought over and made to work the fields.

To understand this history of racialised labour, one could very well tell the story as though the islands and the North 'Queensland' region existed in floating space. Put differently, one could simply tell the story of the region as that of two green spaces, where the blue in-between occupies a rather ephemeral spot in the narrative of 'here' as little more than a material and narrative conduit. Yet, the blue is an integral part of the story. The ocean and rivers of the lands were invested with meaning and came to inform a settler memory, one of capital growth and outward exploration, not one of a racialised slave trade. The waters between the South Sea Islands and 'Australia' are encoded as planes of movement but everything from the naming of the sea(s) to the use of them as places of racial violence firmly emplace them in the stories we tell of place. The waters of the lands – frequently named for and/or by those who invaded – helped to sustain agricultural practices and those that depended on a normalised violence against racialised non-white bodies. To speak of the story of this place, then, one would be remiss to ignore the complex entanglement of the blue with the green, constitutive of each other in the shared (con/de)struction of place.

When many people think of watery spaces, they think of them as reservoirs (to use a 'watery signifier') of marine life and perhaps the consequences of human capitalist hubris. Yet, when we consider how watery spaces can and do have, encoded in them, a memory and a lost history (by way of memory's concern with a strategic myopia), we need to recognise how watery spaces exist around us, through our homes while surrounding them, all the while central in the process of place-making and influencing the memories of our place. Watery spaces are no less powerful natural environs in the management of memory and identity than green spaces, equally imbued as they are with the politics and relations of power that green spaces are.

In education, and perhaps other domains, we owe it to those (including, but not limited to, students) making sense of place to more consciously consider the cyan hues of their place and appreciate that place is not a result of blue or green spaces. Rather, place exists along a spectrum, one that moves and flows just as the rivers and tides do. In those contexts where hegemonic place is forged from ongoing white colonial violence, this work involves critically unpacking the modes of place-making and memorial writing (in place) that celebrates a history of both epistemological and material cyan theft. The community that I live in powerfully anchors its history in the marine spaces, writing the history of naval exploration and racialised violence into the language and measures of place, evidenced by the prominence of figures such as Mathew Flinders and Captains Owen Stanley, Charles Sturt, and John Lort Stokes as namesakes of major thoroughfares through the middle of my hometown. The blue is inescapable, then, however much thinking about place may be 'pulled back to shore'. What education needs is a better recognition of how cyan entanglements make place. Put differently, education thinking needs to better consider cyan hued place-making that critically explores how white memory glorifies invasion and does so through the green and blue spaces of place. A key part of this necessarily involves reframing how place is often quickly conflated with the green as a disruptive endeavour, one that (re-)asserts the blue to create an epistemology of place that is tinted cyan. If water gives life to lands and peoples, it needs to give life to pedagogical work on place and the language that mediates how we understand it. Moreover, we need to tint our critical encounters with placial memory a shade of cyan, one that teases out the complex entanglements of the green and blue as spaces of memorial production and representation.

Terra Australis – the southern land – is a toponym that haunts the history of this place all the while reminding us that invasion was and is a pursuit of land. This is indeed true, but a preoccupation with the 'terra' of this past neglects a recognition that this place was and is *Terra-Aqueous Australis* for invaders, one where their geographic imaginaries shape(d) place through control of cyan spaces. Only when we come to recognise the complex entanglements of cyan spaces can the full extent of this place be properly engaged.

References

Abel, M., Umanath, S., Wertsch, J., & Roediger, H. (2018). Collective memory: How groups remember their past. In M. Meade, C. Harris, P. Van Bergen, J. Sutton, & A. Barnier (Eds.), *Collaborative Remembering: Theories, Research, and Applications* (First edition, pp. 280–296). Oxford University Press.

Alderman, D., & Rose-Redwood, R. (2020). The classroom as 'toponymic workspace': Towards a critical pedagogy of campus place renaming. *Journal of Geography in Higher Education, 44*(1), 124–141. https://doi.org/10.1080/03098265.2019.1695108

Anderson, J. (2010). *Understanding Cultural Geography: Places and Traces*. Routledge.

Azaryahu, M. (2009). Street names and iconography. In R. Kitchin & N. Thrift (Eds.), *International Encyclopedia of Human Geography* (pp. 460–465). Elsevier.

Bashford, A. (2000). 'Is White Australia possible?' Race, colonialism and tropical medicine. *Ethnic and Racial Studies, 23*(2), 248–271. https://doi.org/10.1080/014198700329042

Bonds, A. (2020). Race and ethnicity II: White women and the possessive geographies of white supremacy. *Progress in Human Geography, 44*(4), 778–788. https://doi.org/10.1177/0309132519863479

Butler, A., & Sinclair, K. (2020). Place matters: A critical review of place inquiry and spatial methods in education research. *Review of Research in Education, 44*(1), 64–96. https://doi.org/10.3102/0091732X20903303

Casey, E. (1997). *The Fate of Place: A Philosophical History*. University of California Press.

Casey, E. (2001). Between geography and philosophy: What does it mean to be in the place-world? *Annals of the Association of American Geographers, 91*(4), 683–693. https://doi.org/10.1111/0004-5608.00266

Cook, J. (1893). *Captain Cook's Journal During his First Voyage Round the World Made in H.M. Bark 'Endeavour' 1768–71* (W. Wharton, Ed.). Elliot Stock. (Original work published 1771.)

Curthoys, A. (1999). Expulsion, exodus and exile in white Australian historical mythology. *Journal of Australian Studies, 23*(61), 1–19. https://doi.org/10.1080/14443059909387469

Gieryn, T. (2000). A space for place in sociology. *Annual Review of Sociology*, *26*, 463–496.

Greenwood, D. (2013). A critical theory of place-conscious education. In R. Stevenson, M. Brody, J. Dillon, & A. Wals (Eds.), *International Handbook of Research on Environmental Education* (pp. 93–100). Routledge.

Gruenewald, D. (2008). Place-based education: Grounding culturally responsive teaching in geographical diversity. In D. Gruenewald & G. Smith (Eds.), *Place-Based Education in the Global Age: Local Diversity* (pp. 137–153). Lawrence Erlbaum Associates.

Harley, J. B. (1992). Rereading the maps of the Columbian encounter. *Annals of the Association of American Geographers*, *82*(3), 522–536.

Hercus, L., Hodges, F., & Simpson, J. (Eds.). (2002). *The Land is a Map: Placenames of Indigenous Origin in Australia*. Pandanus Books.

Ho, Y., & Chang, D. (2022). To whom does this place belong? Whiteness and diversity in outdoor recreation and education. *Annals of Leisure Research*, *25*(5), 569–582. https://doi.org/10.1080/11745398.2020.1859389

Israel, A. (2012). Putting geography education into place: What geography educators can learn from place-based education, and vice versa. *Journal of Geography*, *111*(2), 76–81. https://doi.org/10.1080/00221341.2011.583264

Keys Young. (1998). *Homelessness in the Aboriginal and Torres Strait Islander context and its possible implications for the Supported Accommodation Assistance Program*. Department of Family and Community Services. https://webarchive.nla.gov.au/awa/20060830055251/http://facs.gov.au/internet/facsinternet.nsf/VIA/keys/$File/KeysYoung.pdf

Mar, T. (2010). Carving wilderness: Queensland's National Parks and the unsettling of emptied lands, 1890–1910. In T. Mar & P. Edmonds (Eds.), *Making Settler Colonial Space: Perspectives on Race, Place and Identity* (pp. 73–94). Palgrave Macmillan.

Moodie, N. (2018). Decolonising race theory: Place, survivance and sovereignty. In G. Vass, J. Maxwell, S. Rudolph, & K. N. Gulson (Eds.), *The Relationality of Race in Education Research* (pp. 33–46). Routledge.

Moreton-Robinson, A. (2015). *The White Possessive: Property, Power, and Indigenous Sovereignty*. University of Minnesota Press.

Porter, L. (2018). From an urban country to urban Country: Confronting the cult of denial in Australian cities. *Australian Geographer*, *49*(2), 239–246. https://doi.org/10.1080/00049182.2018.1456301

Robinson, S., & Wilson, E. (2008). Preserving the traditions of a 'great race': Youth and national character in Queensland, 1859–1918. *Journal of the Royal Australian Historical Society*, *94*(2), 166–185.

Roediger, H. L., & Wertsch, J. V. (2008). Creating a new discipline of memory studies. *Memory Studies*, *1*(1), 9–22. https://doi.org/10.1177/1750698007083884

Rose-Redwood, R., & Alderman, D. (2011). Critical interventions in political toponymy. *ACME: An International E-Journal for Critical Geographies*, *10*(1), 1–6.

Sanger, M. (1997). Viewpoint: Sense of place and education. *The Journal of Environmental Education*, *29*(1), 4–8. https://doi.org/10.1080/00958969709599101

Sepulveda, C. (2018). Our sacred waters: Theorizing Kuuyam as a decolonial possibility. *Decolonization: Indigeneity, Education & Society*, *7*(1), 40–58.

Smith, B. (2018). Engaging geography at every street corner: Using place-names as critical heuristic in social studies. *The Social Studies*, *109*(2), 112–124. https://doi.org/10.1080/00377996.2018.1460569

Smith, B. (2020). The disciplined winds blow in from the West: The forgotten epistemic inheritance of historical thinking. *Historical Encounters: A Journal of Historical Consciousness, Historical Cultures, and History Education*, *7*(3), 21–32. https://doi.org/10.52289/hej7.300

Smith, B. (2024). Walking the stories of colonial ghosts: A method of/against the geographically mundane. *Journal of Curriculum and Pedagogy*, *21*(1), 84–108. https://doi.org/10.1080/15505170.2022.2098208

Smith, G. (2002). Place-based education: Learning to be where we are. *Phi Delta Kappan, 83*(8), 584–594. https://doi.org/10.1177/003172170208300806
Stanley, T. (2009). The banality of colonialism encountering artifacts of genocide and white supremacy in Vancouver today. In S. Steinberg (Ed.), *Diversity and Multiculturalism: A Reader* (pp. 143–159). Peter Lang Publishing, Inc.
Stickney, J. (2020). Philosophical walks as place-based environmental education. *Journal of Philosophy of Education, 54*(4), 1071–1086. https://doi.org/10.1111/1467-9752.12469
Tuck, E., McKenzie, M., & McCoy, K. (2014). Land education: Indigenous, post-colonial, and decolonizing perspectives on place and environmental education research. *Environmental Education Research, 20*(1), 1–23. https://doi.org/10.1080/13504622.2013.877708
Tucker, B., & Rose-Redwood, R. (2015). Decolonizing the map? Toponymic politics and the rescaling of the Salish Sea. *The Canadian Geographer, 59*(2), 194–206. https://doi.org/10.1111/cag.12140
Tuhiwai Smith, L. (1999). *Decolonizing Methodologies: Research and Indigenous Peoples*. Room 400.
Veracini, L. (2017). Introduction: Settler colonialism as a distinct mode of domination. In E. Cavanagh & L. Veracini (Eds.), *The Routledge Handbook of the History of Settler Colonialism* (pp. 1–8). Routledge.
Watson, I. (2002). Buried alive. *Law and Critique, 13*, 253–269. https://doi.org/10.1023/A:1021248403613
Watson, I. (2009). Sovereign spaces, caring for country, and the homeless position of Aboriginal peoples. *South Atlantic Quarterly, 108*(1), 27–51. https://doi.org/10.1215/00382876-2008-021
Wertsch, J. (2004). *Voices of Collective Remembering*. Cambridge University Press.
Wilkinson, M. (2014). The concept of the absent curriculum: The case of the Muslim contribution and the English National Curriculum for history. *Journal of Curriculum Studies, 46*(4), 419–440.
Willinsky, J. (1998). *Learning to Divide the World: Education at Empire's End*. University of Minnesota Press.
Yazzie, M., & Baldy, C. (2018). Introduction: Indigenous peoples and the politics of water. *Decolonization: Indigeneity, Education & Society, 7*(1), 1–18.

15

ECO-ART AND REELING IN ANTHROPOGENIC ADVERSITY

Robyn Glade-Wright

The state of Queensland, on the northeast coast of Australia, attracts international and national visitors who are drawn to the beauty of a natural wonder of the world, the Great Barrier Reef. The Queensland coast, fringed by the Great Barrier Reef and the Coral Sea, is frequently depicted in picture postcards with highly saturated images replete with brightly coloured tropical corals and marine life. Beach views are portrayed with idyllic piercing blue skies, palm-fringed golden sand, and a jewel-like azure sea.

When I walk along the tideline on the shore of a beach in Far North Queensland, I am delighted to see a red glistening form in the sand. I imagine it could be a piece of coral or a shell of an exotic marine creature. My excitement is shattered when the regular mechanical markings reveal the piece to be a plastic shard from a car tail light. This shoreline, representative of numerous beaches in Queensland, was regarded by many people to be a pristine, cleansing, renewing, and relaxing environment. The area is now sullied by debris from our human society and as a result, my relationships with coastal environments have changed. Seventy to ninety-nine per cent of the plastic that enters the oceans is believed 'to be deposited on shorelines and in marine sediments' (Fulfer & Walsh, 2023, p. 1). Plastic debris that washes ashore is 'estimated to constitute 23% of global ocean plastics' (Meyer et al., 2023, p. 2). In response to Anthropogenic impacts such as plastic pollution that have tainted the 'blue', Eco-artists have 'begun to occupy a new, discursive terrain as agents of public enlightenment' creating works of art to foster sustainable behaviour (Cucuzzella, 2021, p. 2). In this chapter, examples of Eco-art that I have created and exhibited in Queensland are examined to hone

DOI: 10.4324/9781003365501-21

strategies that might aid in the complex task of creating impactful pro-environment messages. I analyse the use of materials, symbolic forms, and emotive content to ascertain their potential contribution to cultivating sustainable behaviours that protect the blue.

Choke

In a series of works called *Choke*, I used discarded waste plastics and rope debris that had washed up on the coast of Far North Queensland (Figure 15.1). Ocean currents and wind distribute the litter disproportionally. Beaches inundated with litter resemble landfill sites, a vision that is a far cry from the imagined beauty of a pristine coast. The *Choke* necklaces are made from discarded plastic floats, children's toys, and plastic rope collected by a marine clean-up organisation along with litter I gathered while walking by the sea. The title of this series reminds people of the destructive and at times fatal impacts of marine creatures choking by entanglement in or ingestion of plastic.

The pieces in the *Choke* series are created in wearable necklace forms, in attractive colours designed to resemble beads and pendant necklaces. The *Choke* necklaces have been made in a range of scales from wearable to unusually large. I aimed to make the work accessible and familiar to link everyday objects and practices with unintended and yet dire outcomes of marine pollution. One necklace features a piece of compressed plastic that may have been formed in a port when layers of plastic were compacted by the weight of water. This hybrid plastic substrate is formed through compression like a sedimentary rock; however, in this case, the material is synthetic Anthropogenic debris. It signals a change in the composition of materials in the environment and challenges concepts I have held about the nature of the landscape. Another necklace is composed of fishing floats and is constructed in a form to suggest a breastplate. In Indigenous cultures, breastplates were used to signify a social role in the community. Breastplates in other cultures were used to identify detainees. In my piece, I imagined the breastplate might suggest a label indicating our collective shame for fouling the blue.

The *Choke* series was exhibited in coastal towns in Queensland, creating a connection between a local environmental issue, the work, and residents and visitors to the region. Exhibitions of art are associated with aesthetic pleasures or entertainment and provide an inviting space for people to consider local environmental impacts in their own time (Sommer & Klöckner, 2021). This counters the tendency for climate change to be viewed as temporally, geographically, and personally distant and brings the demise of the blue closer to home. These works of Eco-art can speak to people who may have diverse levels of commitment to sustainability (Roosen et al., 2018; Sommer et al., 2019;).

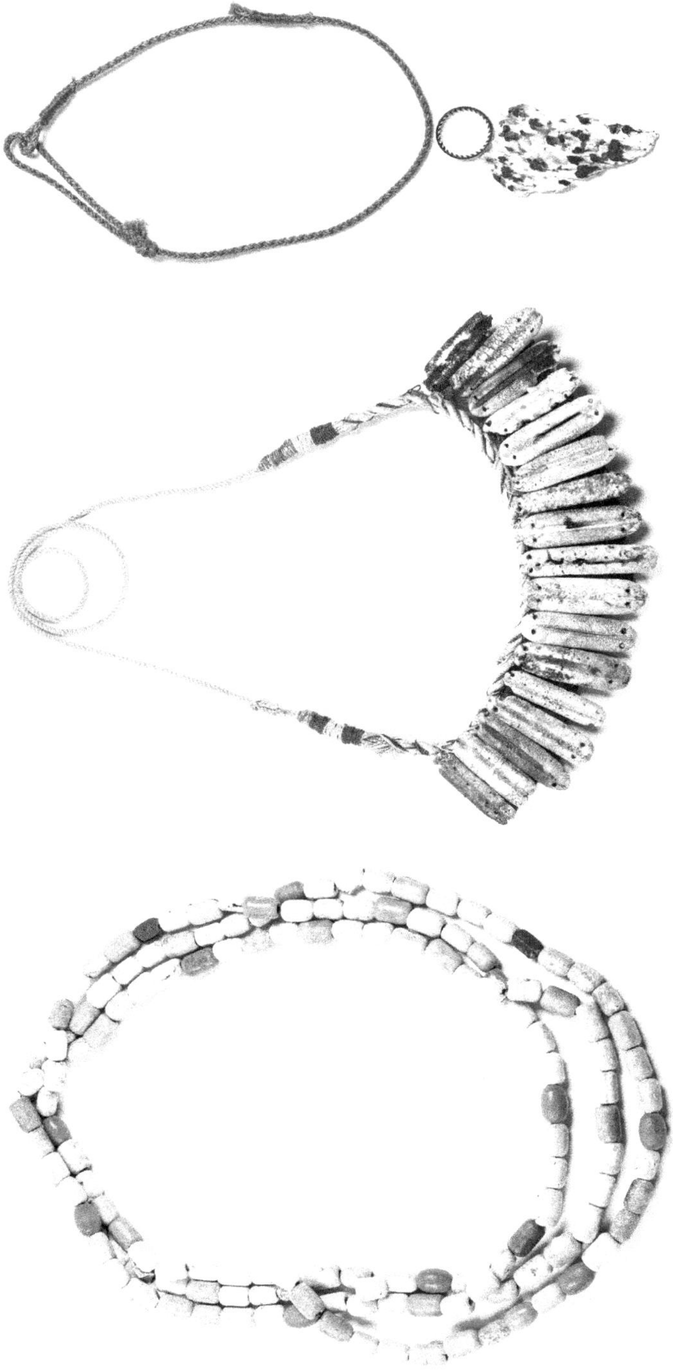

FIGURE 15.1 Robyn Glade-Wright. *Choke Series* 2018. 400 cm x 100 cm x 5 cm. Found plastic floats and rope washed up on Far North Queensland beaches.

Source: Photo: Michael Marzik.

Lampton's Soup

Charlton-Howard et al. remind us that 'Plastics and climate change are intrinsically linked with plastic production currently contributing to 4.5% of global greenhouse gas emissions, exacerbating current damage to our global environments' (2023, p. 2). Crude oil that takes millions of years to form in the ground can be pumped to the surface and shipped to refineries in just a few days (Sojinu & Ejeromedoghene, 2019), and can then be refined and transported to factories and made into items such as plastic knives, forks, and straws in a few more days (Jefferson, 2019). Once created, 50 per cent of all plastic products is used once before they are discarded. Despite their temporary usage, petrochemical plastics are virtually indestructible, as they do not decompose but break up into smaller and smaller pieces over 500 to 1,000 years, and pollute the land, air, and sea (Chamas et al., 2020). Rivers wash approximately 0.5 million tons of plastic per year into the sea, with around 80 per cent of people living in contaminated river basins from macroplastics (Strokal et al., 2023). These plastic particles are joined by other microplastics – from objects like car tyres and synthetic clothing – swirling in huge columns in the ocean in an underwater smog or falling to the sea floor (Li et al., 2020, p. 7). Microplastics in the oceans attract toxic chemical hitchhikers that can enter the food chain, arriving in food sold for human consumption (Cox et al., 2019). People are ingesting approximately 5 grams of plastic each week and over an average life expectancy of 79 years, 'humans consume about 20 kg of plastic' (Myszograj, 2020, p. 203).

To broadcast the sullying of the 'blue' and microplastic pollution in our food I created *Lampton's Soup* 2019 (Figure 15.2). *Lampton's Soup* quotes Andy Warhol's famous *Campbells Soup Cans* (1962), however in my digital images I have listed the number of microplastics contained in each can of soup on the front label. *Lampton's Soup* tangibly reveals the presence of microplastic pollution in our food in local shops and in our pantries. The work demonstrates how Eco-art can deliver perturbing images of environmental pollution in novel ways. Cucuzzella notes the value of Eco-art in enabling 'viewers to develop disturbing images of the often-abstract anthropocentric manifestations of environmental destruction' (2021, p. 13). Roosen and Klöckner (2020) argue that Eco-artists should not be afraid to trigger sadness, helplessness, guilt, and anger in their works of art, as they have the most impact on changing people's intentions and behaviours. During exhibitions of *Lampton's Soup*, audience members expressed amazement, concern, and anger regarding microplastics in the food we eat. The presence of microplastics in the ocean was relatively common knowledge amongst the audience members; however, the thought that our food is polluted was revelatory, perturbing, and upsetting. Epstein (1994) has argued that experiential knowledge gained from viewing art is often more compelling and more likely to influence behaviour than abstract knowledge. This supports Gibson's claim that art-based communication is enlightening, mobilising, and deepens 'what we already know and what we already feel' (2008, p. 585). These capacities of mobilising and

FIGURE 15.2 Robyn Glade-Wright. *Lampton's Soup* 2019. 12 Digital Images 120cm x 160 cm.

deepening what we feel, and the value of experiential knowledge in influencing behaviour, underscore the potential of Eco-art to 'enhance our understanding by providing us with images and perspectives through which experiences may be seen afresh' (Graham, 2005, p. 112). Eco-art can 'normalize concern for the environment, taking it from the realm of "problem" to the realm of general conversation or even entertainment' (Curtis, 2011, p. 190), leading to potential dialogue and even the generation of solutions.

Eco-artists including myself are often motivated and perturbed by scientific data and predictions. During the creative journey, the data and predictions are recast and transformed into fictive imaginative visions. For example, in *Lampton's Soup*, the number of microplastics I listed on the front label was invented. While this practice of fabrication would be anathema in the sciences, the use of 'fiction' is a virtue in the arts. This fictive element of art might explain why art can simultaneously stimulate people's emotions and imagination. This rapid stimulation can lead to a greater understanding of life experiences. The fictive power of art and its alluring, intriguing, and beautiful qualities can amplify the ramifications of scientific data and predictions and present them in a salient manner to reveal new ways of viewing situations and circumstances. As Graham notes, 'whatever we learn from artists is not what we learn from the laboratory' (2005, p. 48). Indeed, as John Bellamy Foster wrote, 'Science and art are two of the preeminent fields of inquiry into nature, with each operating according to its own distinct principles' (2016, para 3). When the importance of climate science is delivered in fictive and novel works of Eco-art, the virtues of both disciplines can combine to help individuals to increase their resolve to protect the blue.

Over-Consumption

Works of Eco-art can enable complex ideas about the ecological crises to be summarised and explored, depicting ecological issues in a tangible form with aesthetic and emotive appeal. The drawing together of complex abstract ideas in a visual form is evident in *Over-Consumption* 2019 (Figure 15.3). This work of art was hung from a gallery ceiling, suggesting the container-laden cargo ship was sailing through a bleached coral reef. *Over-Consumption* with its payload of containers is drawn into direct contact with coral bleaching, which is a result of global warming. Greenhouse gases have accumulated in the atmosphere at an accelerating rate, outpacing the Earth's capacity to adapt to and absorb the excess (Wang et al., 2023). The gases warm the air and sea, and this warming of sea surface ocean temperatures is predicted to increase 'by 2.0–4.8°C by the end of the 21st' century (Habary et al., 2017, p. 566). The warming sea threatens the viability of reef corals. In the past decade, a third of the Great Barrier Reef formed over the past 17,000 to 25,000 years has quietly bleached and died (Bellwood et al., 2019). These once-thriving ecosystems, with vibrant corals and tropical marine life, are now bone-grey calcium graveyards.

FIGURE 15.3 Robyn Glade-Wright. *Over-Consumption* 2019. Cardboard, bamboo, vegetation, paint, glue. 270 cm x 80 cm x 100 cm.

Source: Photo: Michael Marzik.

Cargo ships frequently navigate a passage between the coral cays within the park limits of the World Heritage-listed Great Barrier Marine Park on the northeast coast of Australia. In financial terms, importing consumer goods to Australia by sea is cost-effective, resulting in 11,000 movements of large commercial ships transiting through the marine park annually despite the risk of collisions damaging the reef (Kroon et al., 2020, p. 5). Ehrlich (2010, p. 1) writes: 'Overconsumption by the rich is central to the deterioration of human life-support systems, but is ignored because most business economists, corporate executives, and politicians view it as an unalloyed good'. *Over-Consumption* summarises the causes and outcomes of coral bleaching by depicting the demise of coral reef, in concert with a ship laden with consumer goods. This fabricated scenario sends a call to us all to think about environmentally destructive social and cultural systems, what we buy and use, and how this may impact the environment. As Auer reports, 'aesthetic experience offers insights that are frequently outside the scope of typical scientific inquiry', due to their capacity to draw ideas together and depict consequences (2019, p. 2). In this way, Eco-artists 'move beyond discipline-specific approaches to address complex problems' (Appel & Kim-Appel, 2018, p. 62). When environmental information is delivered in novel works of Eco-art, the virtues of environmental fact and visual art combine and may help individuals to increase their resolve to protect the environment.

Works of Eco-art are frequently made by hand and at a relatable human scale. Eco-artists present personalised stories, contributing their insights and labour for the greater good of the community. British art theorist R. G. Collingwood notes it is not only what *I* feel but what *we* feel that artists, poets, and writers can convey (cited in Graham, 2005, p. 43), indicating that Eco-artists undertake their labour for the benefit of humanity. There is a common view that artists are motivated by emotion; however, Collingwood (1938) argues that artists are motivated by a psychic disturbance that activates one's soul. A psychic disturbance can be positive, like the first blush of love, or negative, such as the emotional trauma that may come from learning the 'current extinction rates are 1,000 times higher than natural background rates of extinction and future rates are likely to be 10,000 times higher' (De Vos et al., 2015, p. 4). Collingwood (1938) claims artists come to understand their underlying emotions through the process of creating a work of art. When I channel concern about plastic pollution into works of Eco-art, I come to understand my deep reverence for the beauty of the world.

And No Eggs Hatched

Like my fellow Eco-artists, I find scientific predictions for the future of the planet to be disturbing, especially given that more than fifty years ago scientists foretold it was unlikely any marine birds would remain uncontaminated by synthetic pollutants (Ohlendorf et al., 1978). It is lamentable that petrochemical plastics production was not curtailed half a century ago. At the time of writing, more than 90 per cent of marine bird species globally have been impacted by plastic and sadly the number is rising, with debris predicted to be found in 99.8 per cent of species by 2050 (Wilcox et al., 2015, p. 11900).

FIGURE 15.4 Robyn Glade-Wright. *And No Eggs Hatched* 2020. 8 cm x 5 cm x 5 cm. Resin, plastic waste.

To process my angst, disappointment, and sorrow regarding the negative impacts of plastic pollution on marine birds, and in the hope of curtailing further harm, I created *And No Eggs Hatched* in 2020 (Figure 15.4). The work consists of life-size eggs made from clear resin, with yolks made from colourful small pieces of plastic waste collected from a local Far North Queensland beach and a supermarket car park. These small eggs with plastic waste yolks would not be viable and no birds would hatch. As with most works of art, *And No Eggs Hatched* is not a didactic factual work. It was inspired by environmental science and takes the data on a creative imaginary journey. Although I am the creator of the art, the piece still generates a strong emotional response in me. It saddens me to imagine that Anthropogenic waste could cause the demise of living creatures including birds. The piece typifies how Eco-artists can respond to the implications of scientific data and present them in a novel form to elicit emotional and cognitive responses. The transdisciplinary alliance between science and works of Eco-art has been shown to result in artefacts that can arouse people's emotions, engage their imagination, and generate an understanding of Anthropogenic causes (Glade-Wright, 2020; Wolf & Moser, 2011).

Microplastics Found in Human Embryo

Eco-art can be a conduit for environmental communication due to its capacity to translate complex abstract scientific data and predictions into visual metaphors.

For example, statements describing the likelihood of a rise of two degrees in global temperatures over the next two decades represent climate change with abstract data. It can be difficult for people to make connections between abstract data, their lifestyle choices, and negative environmental impacts (Wolf & Moser, 2011). Environmental messaging rendered in works of Eco-art can supplement scientific reporting by synthesising complex subjects and presenting them in succinct, coherent, and emotive forms (Sommer et al., 2019). Works of art that generate negative emotions including guilt have been shown to 'significantly contribute to explaining pro-environmental intentions and behaviour' (Keller et al., 2020, p. 266).

The use of a disturbing message designed to elicit negative emotions was evident in my piece *Microplastics Found in Human Embryo* 2018 (Figure 15.5). My aim in creating the piece was to make the most disturbing work I could imagine that would communicate the pressing issue of microplastic pollution. I thought the most distressing image I could create would involve the microplastics threatening the life of an embryo. I made *Microplastics Found in Human Embryo* from 1,000 recycled plastic bottles collected from local accommodation venues. I hand-painted each bottle, using skin-coloured tones to depict the embryo and shades of green, turquoise, and blue for the womb-shaped pool surrounding the embryo. The work was installed on the outside of a building, and it was visible to passing traffic.

To better understand if the aim of *Microplastics Found in Human Embryo* had been successful in highlighting the presence and danger of microplastics, I surveyed (with ethical approval) people who had viewed the piece.[1] In the survey, the emotional impact of *Microplastics Found in Human Embryo* was prominent, with respondents reporting that the work was 'challenging and confronting', 'disturbing', 'thought-provoking', and delivered a 'powerful message'. One of the twenty survey participants wrote: 'It made me feel physically nauseous. It makes me think a lot about plastic use'. The impact of Eco-art on raising attention and issue awareness was acknowledged, with one respondent commenting: 'the emotional impact of the work rekindled a fading concern for the issue'. Behavioural change was mentioned and illustrated by the following participant's comment: 'I think about microplastics more when shopping. It has affected how I shop'. The artwork's emotional impact and its influence on behaviour were underlined by an acknowledgment of the urgency of the issue: 'My understanding of the urgency to address the issues of microplastics was heightened'. Ninety per cent of respondents mentioned beauty, and one respondent commented, 'Glittering in the morning sun is [a] beautiful, but disturbing pollution message'. The frequent use of the word beauty indicates that an aesthetically pleasing form of Eco-art is a suitable 'carrier' of uncomfortable environmental messages (Glade-Wright, 2020).

When I created *Microplastics Found in Human Embryo*, I intended the work to be fictitious; the suggestion that human embryos are impacted by microplastics was creative licence intended to produce negative emotions. However, I later

FIGURE 15.5 Robyn Glade-Wright. *Microplastics Found in Human Embryo* 2018. One thousand recycled bottles, cable ties, spray paint. 350 cm x 540 cm x 10 cm.

Source: Photo: Michael Marzik.

discovered, to my dismay, that microplastics can cross the blood–brain barrier and the placenta, causing cell death and harm (Vethaak & Leslie, 2016). This surprising finding demonstrates how Eco-art can supplement scientific reporting by synthesising complex topics and presenting them in tangible and relatable artefacts (Sommer et al., 2019). Works of Eco-art can communicate the unfolding horror of plastic pollution using readily understood images to deliver confrontational messages that engage a broad cross-section of the community.

Conclusion

In summary, the works of Eco-art that I have made and exhibited in Far North Queensland over the past decade have been inspired by data and predictions generated by environmental scientists and my experience of changes to the coastline. To communicate the negative impacts of a changing climate and environmental pollution, I have used tangible images that relate to the lives of the people in the coastal region of Queensland. Images of cargo ships, bleached coral, foodstuff, and jewellery are used to transmit environmental messages in a form that relates to many people. My works of Eco-art deliver cognitive and emotive messages in an attractive form. The works sit quietly in the gallery; however, the use of materials, symbols, and metaphors speak of Anthropogenic carnage that negatively impacts marine birds, animals, corals, as well as food supply and human health. This reflection of the impact of Eco-art indicates that such work can synthesise complex topics and present them in tangible artefacts that play a role in motivating sustainable behaviour by engaging people's emotions, imagination, and intellect. Eco-artists can contribute to promoting sustainable behaviour to reel in Anthropogenic adversity and protect the blue.

Note

1 This data collection was conducted under ethics approval from James Cook University (reference: H7741).

References

Appel, J., & Kim-Appel, D. (2018). Towards a transdisciplinary view: Innovations in higher education. *International Journal of Teaching and Education*, *6*(2), 61–74. doi: 10.20472/TE.2018.6.2.004

Auer, M. R. (2019). Environmental aesthetics in the age of climate change. *Sustainability*, *11*(18), 5001. https://doi.org/10.3390/su11185001

Bellwood, D. R., Pratchett, M. S., Morrison, T. H., Gurney, G. G., Hughes, T. P., Álvarez-Romero, J. G., Day, J. C., Grantham, R., Grech, A., Hoey, A. S., Jones, G. P., Pandolfi, J. M., Tebbett, S. B., Techera, E., Weeks, R., & Cumming, G. S. (2019). Coral reef conservation in the Anthropocene: Confronting spatial mismatches and prioritizing functions. In *Biological Conservation*. https://doi.org/10.1016/j.biocon.2019.05.056

Chamas, A., Moon, H., Zheng, J., Qiu, Y., Tabassum, T., Jang, J. H., . . . & Suh, S. (2020). Degradation rates of plastics in the environment. *ACS Sustainable Chemistry & Engineering*, *8*(9), 3494–3511. https://pubs.acs.org/doi/10.1021/acssuschemeng.9b06635

Charlton-Howard, H. S., Bond, A. L., Rivers-Auty, J., & Lavers, J. L. (2023). 'Plasticosis': Characterising macro- and microplastic-associated fibrosis in seabird tissues. *Journal of Hazardous Materials*, *450*, 131090. https://doi.org/10.1016/j.jhazmat.2023.131090

Collingwood, R. G. (1938). *The Principles of Art*. Oxford: Clarendon Press.

Cox, K. D., Covernton, G. A., Davies, H. L., Dower, J. F., Juanes, F., & Dudas, S. E. (2019). Human consumption of microplastics. *Environmental Science & Technology*, *53*(12), 7068–7074. https://doi.org/10.1021/acs.est.9b01517

Cucuzzella, C. (2021). Making the invisible visible: Eco-art and design against the anthropocene. *Sustainability*, *13*(7), 3747. https://www.mdpi.com/2071-1050/13/7/3747

Curtis, D. J. (2011). Using the arts to raise awareness and communicate environmental information in the extension context. *Journal of Agricultural Education and Extension*, *17*(2), 181–194. doi: 10.1080/1389224X.2011.544458

De Vos, J. M., Joppa, L. N., Gittleman, J. L., Stephens, P. R., & Pimm, S. L. (2015). Estimating the normal background rate of species extinction. *Conservation Biology*, *29*, 452–462. https://doi.org/10.1111/cobi.12380

Ehrlich, P. R. (2010). The MAHB, the culture gap, and some really inconvenient truths. *PLOS Biology*, *8*(4), e1000330. https://doi.org/10.1371/journal.pbio.1000330

Epstein, S. (1994). Integration of the cognitive and the psychodynamic unconscious. *American Psychologist*, *49*(8), 709–724. https://doi.org/10.1037/0003-066X.49.8.709

Foster, J. B. (2016). Nature. *Monthly Review*, *68*(1). https://monthlyreview.org/2016/05/01/nature/

Fulfer, V. M., & Walsh, J. P. (2023). Extensive estuarine sedimentary storage of plastics from city to sea: Narragansett Bay, Rhode Island, USA. *Scientific Reports*, *13*(1), 1–11. https://doi.org/10.1038/s41598-023-36228-8

Gibson, J. (2008). Cognitivism and the arts. *Philosophy Compass*, *3*(4), 573–589. https://doi.org/10.1111/j.1747-9991.2008.00144.x

Glade-Wright, R. (2020). The persistence of plastic: Environmental public art and microplastic pollution. *PAN: Philosophy Activism Nature*, *15*, 16–26. https://search.informit.org/doi/10.3316/informit.601770458381279

Graham, G. (2005). *Philosophy of the Arts: An Introduction to Aesthetics* (3rd edn). New York: Routledge.

Habary, A., Johansen, J. L., Nay, T. J., Steffensen, J. F., & Rummer, J. L. (2017). Adapt, move or die – how will tropical coral reef fishes cope with ocean warming? *Global Change Biology*, *23*(2), 566–577. https://doi.org/10.1111/gcb.13488

Jefferson, M. (2019). Whither plastics? – Petrochemicals, plastics and sustainability in a garbage-riddled world. *Energy Research & Social Science*, *56*, 101229. https://doi.org/10.1016/j.erss.2019.101229

Keller, A., Sommer, L., Klöckner, C. A., & Hanss, D. (2020). Contextualizing information enhances the experience of environmental art. *Psychology of Aesthetics, Creativity, and the Arts*, *14*(3), 264. http://dx.doi.org/10.1037/aca0000213

Kroon, F. J., Berry, K. L., Brinkman, D. L., Kookana, R., Leusch, F. D., Melvin, S. D., . . . & Williams, M. (2020). Sources, presence and potential effects of contaminants of emerging concern in the marine environments of the Great Barrier Reef and Torres Strait, Australia. *Science of the Total Environment*, *719*, 135140.

Li, Y., Zhang, H., & Tang, C. (2020). A review of possible pathways of marine microplastics transport in the ocean. *Anthropocene Coasts*, *3*(1), 6–13.

Meyer, A. N., Lutz, B., & Bergmann, M. (2023). Where does Arctic beach debris come from? Analyzing debris composition and provenance on Svalbard aided by citizen scientists. *Frontiers in Marine Science*, *10*, 1092939. https://doi.org/10.3389/fmars.2023.1092939

Myszograj, M. (2020). Microplastic in food and drinking water – environmental monitoring data. *Civil and Environmental Engineering Reports*, *30*(4), 201–209.

Ohlendorf, H. M., Risebrough, R. W., & Vermeer, K. (1978). Exposure of marine birds to environmental pollutants. Retrieved from https://archive.org/details/exposureofmarine00ohlerich

Roosen, L. J., & Klöckner, C. A. (2020). Art and documentaries in climate communication: Experiencing the reality of climate change and leading the way to change. *Art Research International*, *5*(2). https://journals.library.ualberta.ca/ari/index.php/ari/article/view/29520

Roosen, L. J., Klöckner, C. A., & Swim, J. K. (2018). Visual art as a way to communicate climate change: A psychological perspective on climate change-related art. *World Art*, *8*(1), 85–110. https://doi.org/10.1080/21500894.2017.1375002

Sojinu, S. O., & Ejeromedoghene, O. (2019). Environmental challenges associated with processing of heavy crude oils. *Processing of Heavy Crude Oils*, *241*.

Sommer, L. K., & Klöckner, C. A. (2021). Does activist art have the capacity to raise awareness in audiences? – A study on climate change art at the ArtCOP21 event in Paris. *Psychology of Aesthetics, Creativity, and the Arts*, *15*(1), 60. http://dx.doi.org/10.1037/aca0000247

Sommer, L. K., Swim, J. K., Keller, A., & Klöckner, C. A. (2019). 'Pollution pods': The merging of art and psychology to engage the public in climate change. *Global Environmental Change*, *59*, 101992. https://doi.org/10.1016/j.gloenvcha.2019.101992

Strokal, M., Vriend, P., Bak, M. P., Kroeze, C., van Wijnen, J., & van Emmerik, T. (2023). River export of macro- and microplastics to seas by sources worldwide. *Nature Communications*, *14*(4842). https://doi.org/10.1038/s41467-023-40501-9

Vethaak, A. D., & Leslie, H. A. (2016). Plastic debris is a human health issue. *Environmental Science & Technology*, *50*, 6825–6826. https://doi.org/10.1021/acs.est.6b02569

Wang, S., Foster, A., Lenz, E. A., Kessler, J. D., Stroeve, J. C., Anderson, L. O., . . . & Hausfather, Z. (2023). Mechanisms and impacts of Earth system tipping elements. *Reviews of Geophysics*, e2021RG000757.

Wilcox, C., Van Sebille, E., & Hardesty, B. D. (2015). Plastic in seabirds is pervasive and increasing. *Proceedings of the National Academy of Science*, *112*(38), 11899–11904. https://doi.org/10.1073/pnas.1502108112

Wolf, J., & Moser, S. C. (2011). Individual understandings, perceptions, and engagement with climate change: Insights from in-depth studies across the world. *Wiley Interdisciplinary Reviews: Climate Change*, *2*(4), 547–569. https://doi.org/10.1002/wcc.120

16

WAVES OF COGNITION

Towards an Australian Blue Shakespeare ecosystem

Alys Daroy, Joshua Zeunert, and Rahul K. Gairola

Introduction: Shakespeare's global currents Down Under

Australia, as an 'island girt in with the ocean' (*Henry VI Part 3*, 4.8.20), is at once connected to yet separated from other continents by flows and currents.[1] Australia's blue ecosystems are filled with diverse organisms, habitats, and life-webs that together comprise a complex and dynamic whole. Similarly, the blue humanities is generated and sustained by interdisciplinary interactions and underpinned by ecological systems thinking, which considers the interrelationships and interdependencies. With these blue entanglements in mind, this chapter proposes an initial foray into Australian Blue Shakespeare's future possibilities. Our intention is twofold. Firstly, we argue that blue literary, theatrical, and pedagogical methods must seek to emulate multi-scalar ecosystems rather than privileging individual foci and distinct fields as corresponds with blue ecological systems' interconnectedness. We note that this blue ecosystem includes climatic and aquatic realms, their more-than-human inhabitants, the human brain (80 per cent water) (Nichols, 2014) and body (70 per cent water) (Mentz, 2019, p. 383), as well as cultural conceptions of blue environments.

Secondly, we propose three perspectives in Australian Blue Shakespeare in order to canvas these material and conceptual elements: 1) ecological cognition; 2) landscape studies and ecology; and 3) Australian postcolonial critique, as represented by the authors' fields. Ecological cognition, or cognitive ecology, views cognition as embodied, extended, embedded, and enacted across mind, body, and environment (Hardy-Vallée & Payette, 2008, pp. 1–2). This offers a way by which to view Shakespeare's blue metaphors and allusions as expressions of ecological selfhood embedded in blue environments. Landscape studies (incorporating both maritime as well as terrestrial aquatic environments) and ecology can better

DOI: 10.4324/9781003365501-22

inform understanding of Australia's diverse and unique environments and improve ecological literacy within literary and artistic interpretations of Shakespeare's work. Shakespearean ecocriticism already envelops intersectional approaches, but expressly Australian postcolonial critique may facilitate greater interrogation of the country's imperialist history and its continued effects, especially through Indigenous and non-Eurocentric voices. This chapter therefore starts what will hopefully develop into a longer conversation about Shakespeare's intersections between mind, body, and blue environments in the Australian context. Beginning with a discussion of *Twelfth Night*'s blue presence/absence, we move to a brief examination of Shakespeare's wider watery worlds through the lens of ecological cognition. We then turn to the place-based potential for Australian blue interpretations, concluding with an interview with postcolonial scholar Rahul K. Gairola in relation to Australian blue pedagogy and *The Tempest*.

Australian Blue Shakespeare

Shakespeare's blue worlds are not static discrete entities but merge and meld with ecological connectivity. Thus, Shakespeare's ocean melds with vaporous clouds (*Julius Caesar*, 1.3.7–8), 'fresh streams meet in one salt sea' (*Henry V*, 2.1.210), and rain falls for the whole 'earth's increase' (*Henry VI Part 2*, 3.2.384). Shakespeare's early modern conceptions of an interconnected organic world have already been extensively examined by such scholars as Randall Martin (2015) and Todd Borlik (2019). More pertinently to the blue humanities, Steve Mentz's (2009, 2018, 2019) and Dan Brayton's (2011, 2012) seminal works have put Shakespeare's ocean on the ecocritical map and birthed a buoyant Blue Shakespearean sector. Within the Australian context, however, Shakespeare's vital, vibrant blue matter presents opportunities for further exploration. Claire Hansen (2019, 2023a, 2023b) is helping to shape the Blue Australian Shakespeare movement with her place-based and wider research. Jennifer Mae Hamilton (2017) examines Shakespeare's blue in relation to meteorological events and the storm trope, though in a global theatrical rather than expressly local context. Other Australian Blue Shakespeare research and practice includes Rob Conkie's (2018) adaptation of *King Lear* and ecological reimagining of the play's meteorological events, and Shakespeare South's ecotheatrical *Twelfth Night* utilising blue physical ecodramaturgy and ecoscenography (Daroy, 2021).

Currently, though, Australian Blue Shakespeare tends to contribute to the broader international conversation rather than examine the continent's unique cultural and material environments. The exception to date is Gretchen Minton's *Salt Waves Fresh* (2023), an ecological adaptation of *Twelfth Night* exploring responses to cyclones, floods, and destruction of the Great Barrier Reef. Minton's ecodramaturgical reinterpretation of the play combines First Nations' histories with Far North Queensland's uncertain blue future. Similarly, Hansen (2023a) draws attention to the fact that examining Shakespeare through the lens of the blue humanities shifts attention not only to material watery spaces but also histories, futures, and new ways of thinking within an age of environmental crisis. We agree, and argue for the development of an Australian Blue Shakespeare methodology grounded in systems-thinking that accounts for the continent's turbulent cultural and environmental histories and precarious futures. To this end, we hope that this discussion functions as a preliminary paddle to what will hopefully lead to later deep dives into interdisciplinary and intersectional Blue Shakespeare methods. Within this brief chapter, we hope to offer an exploration of Shakespeare's 'blue' in Australia that dips between the tangible, ethereal, local, and universal as a departure point for further place-based immersion. We begin with a short examination of the presence/absence of blue environments within *Twelfth Night* as a precursor to the blue's cognitive impact and Shakespeare's potential for the Australian blue humanities.

Blue presence/absence in *Twelfth Night*

The blue is slippery, oozing between mind, body, and environment. Shakespeare's *Twelfth Night* offers an interesting case study of the manner in which blue environments may seem vitally present in a play's atmosphere despite an absence of tangible blue locations and even references. This hints at the blue's cognitive impact and potential for further literary and artistic interventions through which to examine Australia's own blue crises. The following section examines some of the ways in which *Twelfth Night* has been read and staged as a 'blue world' play. Applying a close reading of blue references within Shakespeare's thirty-eight sole and co-authored plays from a comprehensive study (Daroy, 2022), we here argue that contrary to popular conception, the play's blue is conspicuous in its absence rather than its presence. This, we suggest, does not negate its stormy maritime atmosphere but rather invites further questions as to how and why the text invokes such a powerful conception of aquatic environments.

Twelfth Night is heavily associated with water, in particular storms and the sea. Alison Findlay, Liz Oakley-Brown, and Nicola Spelman, for instance, note that 'in the last five years or so, *Twelfth Night's* dramatisation of seascapes as well as environments has been examined' (2014, p. 6, [Frye, 1957]). Northrop Frye earlier incorporates the play within a 'group of "sea" comedies' that includes *The Comedy of Errors*, *Pericles*, and *The Tempest* (1957, pp. 184–185). Mentz also classes *Twelfth Night* as a 'sea-comedy', with Viola's 'alienation' representing its 'relative blue-ness' (2018, p. 258). In reconfiguring Frye's green world into 'dynamic and disorderly forms more compatible with post-sustainability ecological thinking', Mentz considers *Twelfth Night*'s 'redemptive elements' to 'emerge out of the actions of its maritime bit-players' (2018, p. 258). Moving to the blue of rainstorms, Martin highlights the play's 'wind and the rain' (2015, p. 10), while Jalal Khan notes its 'storms, moors and seashores' (2019, p. 96). Productions also frequently emphasise the play's blue environment. Lisa Anderson emphasises the play's historically blue performance environment, including Sir Henry Irving's 1884 production 'beside the blue sea' and the first film adaptation by J. Stuart Blackton in 1901 in which Viola emerges from Long Island's Bay Shore (2014, p. 54). More recently, the Royal Shakespeare Company (RSC) classed the play as part of its Shipwreck Trilogy (Findlay & Oakley-Brown, 2014, p. 6), while Trevor Nunn's 1996 film adaptation by Cornwall's coast features what Peter J. Smith terms a 'maritime obsession' (2014, p. 395).

For a play prefigured by a storm at sea, however, *Twelfth Night*'s text contains fewer references to both precipitation and the ocean than might be expected. There are implied spatterings here and there in the 'heavens rain odours on you!' (3.1.85) and 'the wind and the rain' (5.1.382) of Feste's song.[2] Maritime references fare better, whether metaphorically through 'flood of fortune' (4.3.11), encased in simile 'as hungry as the sea' (2.4.101), in allusions to 'salt waves fresh' (3.4.383), through word play, as in 'Will you hoist sail' (1.5.197), and topographically through recounting the shipwreck prior to the play's action: 'I saw him hold acquaintance with the waves' (1.2.16). To offer some comparison, although *Twelfth Night* is only approximately 62 per cent of *Hamlet*'s length, the latter contains roughly three to four times its meteorological references, though *King Lear* far exceeds both (Daroy, 2022). Based on a systematic analysis of his complete works, *Twelfth Night* in fact proves fairly unremarkable with respect to its blue inclusions. References are approximately equal to thirteen non-blue world plays, with more blue inclusions in thirteen others of mixed genre (Daroy, 2022). Despite its strong associations to blue place and environment, *Twelfth Night* seems to hint, suggest, or gesture rather than to overtly state.

There is something, then, particularly cognitively impactful about *Twelfth Night*'s blue content, a dual presence/absence that is at once intangible yet affective. Before we turn to possible ways to apply Shakespeare's topographical indeterminacy to place-based Australian contexts, we first consider how ecological cognition may help to situate blue environments as flowing through mind, body, and environment. This, we propose, may assist ecological interpretations of his plays by drawing greater attention to our psychological and physical embeddedness within material ecosystems, thus highlighting their importance to human survival, culture, and also selfhood. By understanding our identities as entangled within blue environments, we may hope to find fresh readings and dramaturgical analyses of Shakespeare's plays that emphasise our species' vital link to hydrological systems and the need to preserve and rehabilitate ecosystems in Australian place-based contexts.

Ecological cognition and Shakespeare's blue mind

Though scholars such as Mentz (2019) have already examined the preponderance of water in Shakespeare's plays from an ecological and evolutionary perspective, Shakespeare's blue references have yet to be considered from the standpoint of ecological cognition. We may only briefly dip our toe in the water of this deep stream within this chapter's bounds but do so with a view to further analysis. Ecological cognition views neural systems as shaped by the prevailing environments and conditions of life during *Homo sapiens*' evolutionary history in the Pleistocene era (2.6 million to 11,700 years ago). Water was essential to our species' extensive period of evolution in pre-urbanised environments as a source of hydration, food, wayfinding, transportation, and hygiene. Blue metaphor hence functions as a product of a historic and continuing organism-environment system that is visible in cultural artefacts and enacted within the reader's or audience's mind/body (Gallagher & Lindgren, 2015, p. 392). Shakespeare's metaphoric meaning when viewed in this way may be seen to emerge from evolutionary and continued 'projections of structure between a living organism and its perceived or imagined environment' (Jensen & Greve, 2019, p. 9). In other words, the 'natural world . . . indelibly shape[s] the metaphorical language we use' (Gibbs, 2019, p. 43). In this view, metaphor functions not as a mere literary device but 'an act of discovery' about our connection to environments (Gibbs, 2019, p. 43). Examining Shakespeare's presence and absence of blue allusions more closely offers us an opportunity to take the 'environmental (material and physical) aspects of language' more 'seriously' (Jensen & Greve, 2019, p. 13).

A quick zoom out at Shakespeare's oeuvre reveals just how present the blue world is within the plays. Shakespeare's blue is distributed across plays with settings as diverse as towns ('ocean whelm them all!' (*Merry Wives of Windsor*, 2.2.129), battlefields ("Tis with my mind / As with the tide swell'd' (*Henry VI Part 2*, 2.3.62–63), and courts ('The wild sea of my conscience' (*Henry VIII*, 2.4.198). Within these swirling currents, the blue is active both in its presence and absence ('Great floods have flown / From simple sources, and great seas have dried' (*All's Well That Ends Well*, 2.1.138–139)). Furthermore, Shakespeare's blue is not encased within static metaphor but dynamic and sensory. It murmurs and chafes (*King Lear*, 4.6.20–22), rises and seeps (*Troilus and Cressida*, 1.3.111–113), reflects (*Winter's Tale*, 4.4.174–175), dissolves (*Antony and Cleopatra*, 4.14.10–11), is textured (*Henry VI Part 2*, 3.1.53), and fuses multiple materialities with notions of self (*Merchant of Venice*, 1.1.34). This creates a multilayered sensorium in the evolved blue mind's eye. This cerulean richness offers myriad ecodramaturgical, ecoscenographic, ecocritical, and ecopedagogical possibilities for reinterpretation against the backdrop of the Anthropocene's blue crises. How, though, may they be applied to an Australian context? What follows is a brief foray into what we hope will become a wider critical conversation.

The Australian place context and Blue Shakespeare potential

Australia's ecosystems and blue environments are delicately fragile and ancient, with complex life-webs, biodiversity hotspots, and co-existence of species evolving from aeons of relative stability and isolation. Australia's First Nations peoples transformed the pre-human ecologies over tens of millennia, reshaping the continent into fire-reliant cultural–ecological systems (Harari, 2015; Pyne, 2016). Yet the brutalism of the occupying British post-1788 rapidly decimated the continent's First Nations Peoples (Ma Rhea, 2017) and continuation of extractivist economic primacy and anthropocentric land management equates to ongoing ecological havoc (Creswell et al., 2021; Lines, 1999; Thackway et al., 2018). The Barka/Baaka/Darling's Riverbeds, for example, are now regularly dry when a century ago paddle-steamers chugged along this terrestrial blue vein (SLNSW, 1905). As the driest inhabited continent, contemporary Australia's blue pressures are heightened by perpetual economic and continuing population growth, which reduces per-capita availability of biocapacity and increases pressures on ecosystems and their ability to provision services and assimilate wastes. While the familial El Niño/La Niña Southern Oscillation regularly creates too little rainfall-drought or too great flooding in Australia, ongoing atmospheric greenhouse gases recalibrate prior patterns and exacerbate unpredictability (Flannery, 2005; Lovelock, 2006). This strains agricultural production dependent on steady weather patterns (Zeunert, 2018), as large-scale infrastructures attempt to mediate pressures stemming from unstable natural systems. The resources and agribusiness sectors mine finite fossil groundwater in order to operate in unlikely place-contexts resembling remote moonscapes. Irrigation infrastructures transform arid and semi-arid environments into vast networks of water-intensive crops like cotton and feed for livestock, where single farms occupy greater spatial footprints than the world's largest megacities (Zeunert, 2024).

Furthermore, Australia is one of the world's largest freshwater exporters due to the water required to produce agricultural commodities and the subsequent trade of 72 per cent of the nation's total to overseas markets (ABARES, 2022). Agricultural chemical use is widespread, resulting in eutrophication and runoff into waterways and oceans, impacting, for one, the world's largest ecosystem – the Great Barrier Reef. Past measures in our built environments, such as one in a hundred year floods, are made obsolete by increasingly regular events at unprecedented scales. Concreted and hard environments in (sub)urban expanses are regularly constructed adjoining (or over) creeks, rivers, deltas, and floodplains, and continue their advance amidst neoliberal urban planning akin to revolving doors for commercial developers (Zeunert & Freestone, 2022). Combining Shakespeare's blue sensorium with these omnipresent contemporary challenges offers rich potential to link past, present, and future, as well as intertwined novel and hybrid cultural worlds that collectively need to better steward the continent.

The continuing onslaught on Australia's ecological systems offers us a useful way to view the blue's many manifestations amidst contemporary issues and pressures. There is exciting room for growth within Australia's Blue Shakespeare productions. Ecocritically, ecodramaturgically, and ecoscenographically, Blue Shakespeare analyses can read the plays through specifically Australian challenges and environmental events. *Antony and Cleopatra*'s Nile and references to mud, for instance, may form an interesting ecodramaturgical foil for the ailing Murray–Darling River system (Australia's 'food bowl'), while *King Lear*'s storm offers a rich source from which to examine responses to extreme weather events such as climate change. There are, however, numerous other ways in which to link the plays' impactful blue content to specifically Australian sites, species, and stories, from intertwining verbatim theatre responding to environmental anxiety, catastrophe, and celebration to regenerative stage practices. Additionally, colonial cultural and environmental imposition offers a further angle for the development of Australian pedagogy. The following short interview with Rahul K. Gairola offers a glimpse of potential ways to socioculturally position Blue Shakespeare studies within Australian pedagogical contexts.

Blue Australian pedagogy and *The Tempest*: Interview with Rahul K. Gairola

Has there been any innovation or new knowledge generated through your pedagogical methods in relation to The Tempest *and the blue Australian context?*

I think that the new knowledge to be drawn from teaching Shakespeare in the Australian context comes from the blue historical context that has shaped the play's narrative. What's been really interesting for me is that my students have pressed me to think about Australia, surrounded as it is by water, as a stand-in for the play's isle. Its inhabitants may therefore be viewed as figures that are subversive and transgressive to European ideals of beauty or propriety that have been imported from across the globe's oceans. This is crystalised in the character of Caliban as an Indigenous character, as dramatised by the Sydney Theatre Company's 2022 production (Williams, 2022). My students, for example, have likened questions of Caliban's agency to Australia's upcoming referendum to enshrine an Aboriginal and Torres Strait Islander Voice in the Australian Constitution (Browne, 2022). Australia is a continent whose borders are sculpted, quite literally, by the blue and the blue remains highly significant. (Then) Lieutenant James Cook and subsequent fleets, for example, used the blue as the terrain upon which they sailed to occupy this '*Terra nullius*' for the British Empire. Like Caliban's comparisons to a 'strange fish' (2.1.113), Australia's Aboriginal and Torres Strait Islander people were dehumanised, justifying their dispossession. Moreover, Australia continues to be fashioned by the blue, from the migratory movements of diaspora to the closing and opening of borders due to Covid-19. On top of this, Australia is very much shaped by floods, droughts, and marine environments and inhabitants that continue to be destroyed. Furthermore, just as *The Tempest*'s colonisers are supported by Indigenous local knowledge, current-day Australia must continue to learn from First Nations' knowledge systems and interconnection with land and water.

How do these ideas reflect the play's blue absence and presence?

Australia's blue world negotiates presence and absence in the manner of *Twelfth Night*. The absence of blue within Shakespeare's work is immediately a presence that brings attention to itself. The best analogy that I can make is that just because I do not have a glass of water and I am dying of thirst does not render water absent. Rather, the mere fact that I'm craving a glass of water makes water immediately present through its conspicuous absence. This is a metaphor describing a human body, but it is immediately relevant when we talk about geographical terrain and the bushfires and droughts that have afflicted Australia in recent years. *The Tempest* is a very interesting play in contrast to *Twelfth Night*, which we could say is shaped by a pervasive blue absence. I've argued and spoken about *The Tempest* having a preponderance of blue homelessness. This is particularly evident in Caliban, who at once emits, as I like to put it, verses and curses (Gairola, 2019, p. 40). When viewed through an Australian postcolonial blue lens we can see how the play's colonialism as both historical fact and narrative erupts through the text and demonstrates identity politics from the play's historical moment to today. The persistence and legacy of this identity politics continue to shape the current moment. *The Tempest*'s island is Caliban's, or rather his mother's, the witch Sycorax. Sycorax is mentioned but never appears, so she is also an absent presence throughout the play. Shakespeare's blue world is both gendered and racialised. It is a fluid environment that conveys identity as well as terrain in which conquest is enacted. There is therefore a kind of 'blue homelessness' in *The Tempest* whereby Caliban is both enslaved and homeless on his own island.

What unique pedagogical challenges and/or opportunities are present when teaching The Tempest *in the Australian classroom?*

Teaching Shakespeare in the Australian context opens up unique pedagogical challenges and/or opportunities surrounding ideas of indigeneity, space, place, and the concept of home. The blue humanities offers a lens through which to approach this in the classroom. This is due to the blue world's significance not only to Australia's Indigenous people but to all living on the continent. In my own research, I have examined the way in which the blue world continues to shape migratory diasporic trends and patterns (Gairola, 2016, p.14). At the same time, this includes the more-than-human world. The challenge of teaching the blue humanities in Australia is hence also the opportunity, in that we can continue to make the text relevant in the current historical moment. When we discuss the impact of water, both on culture and ecosystems, we can see how a four-century-old text continues to matter. Though Shakespeare's plays are rooted in a historical moment, they continue to be relevant. We are all connected through oceans, climate patterns, and tides. Although we talk about the Australian blue environment, we can never forget that this is always a part of other waterways that impact and flow throughout the world.

Similarly, water flows through all human beings, living creatures, territories, and environments. We are thus connected in Australia and beyond by biopolitics and biopower but also through our blue bodies. A blue lens can assist us within the Australian classroom to analyse Shakespeare's as well as our own cultural contexts and material environments as both local and global.

Conclusion: Blue ecosystem

Shakespeare's depth and liquidity offer a compelling portal into our planet's interconnected systems and their heuristic ramifications for the Australian blue humanities. The plays' blue metaphors and allusions invite us to question ideas of presence and absence of place and the blue's seepage throughout mind, body, and environment. This is particularly evident in *Twelfth Night*, in which the blue's presence hovers ethereally somewhat out of grasp while simultaneously grounding the play in a maritime context. So too *The Tempest*'s topography translates both historically and ecologically to Australian culture and continent. A lack of systems-knowing of ecological connectedness ultimately leads to actions threatening and undermining Australia's blue environments: from the aquamarine gems like the Great Barrier and Ningaloo Reefs to habitats, drinking water, and the microplastics in our aquatic life (see Robyn Glade-Wright's chapter in this collection), creeks, and agricultural fields flowing into our oceans (and foods). A cross-disciplinary approach to Australian Blue Shakespeare facilitates ecological specificity and cultural relevance to the Anthropocene's intersectional challenges. Nesting these scales and multi-disciplinary lenses within an ecosystem perspective may help to navigate Australia's cultural–ecological challenges. It is our hope that further exploring Shakespeare's blues worlds may offer the Australian blue humanities another fluid device through which to examine and harness texts' and artistic works' compelling blue impact.

Notes

1 All references to Shakespeare's texts in this chapter are from *The Arden Shakespeare Complete Works* (2013), edited by Proudfoot et al.
2 With respect to the freshwater blue, *Twelfth Night* contains only one reference to rivers in 'How runs the stream?' (4.1.60).

References

Anderson, L. (2014). Performance and adaptation. In A. Findlay & L. Oakley-Brown (Eds.), *Twelfth Night: A Critical Reader* (pp. 52–70). Bloomsbury.
Australian Bureau of Agricultural and Resource Economics and Sciences (ABARES). 'Snapshot of Australian Agriculture 2022'. Australian Government Department of Agriculture, Fisheries, and Forestry. Accessed 1 May 2022. https://www.awe.gov.au/abares/products/insights/snapshot-of-australian-agriculture-2022

Borlik, T. (2019). *Literature and Nature in the English Renaissance: An Ecocritical Anthology*. Cambridge University Press.

Brayton, D. (2011). Shakespeare and the global ocean. In L. Bruckner & D. Brayton (Eds.), *Ecocritical Shakespeare* (pp. 173–190). Taylor and Francis.

Brayton, D. (2012). *Shakespeare's Ocean: An Ecocritical Exploration*. University of Virginia Press.

Browne, B. (2022). Referendum machinery submission summary introduction. The Australia Institute, Australia. Retrieved from https://australiainstitute.org.au/wp-content/uploads/2022/12/P1349-Referendum-machinery-submission-Web.pdf

Conkie, R. (2018). Nature's above art (an illustrated guide). *Shakespeare Bulletin*, *36*(3), 391–408. http://doi.org/10.1353/shb.2018.0037

Cresswell, I., Janke, T., & Johnston, E. (2021). *Australia: State of the Environment 2021*. Commonwealth of Australia.

Daroy, A. [Dir.]. (2021). *Botanica Lumina: Twelfth Night*. [Play]. Shakespeare South.

Daroy, A. (2022). *Biophilic Shakespeare: Towards an Ecology of Form*. [Doctoral Thesis]. Monash-Warwick Alliance.

Findlay, A. and Oakley-Brown, A. (Eds.). (2014). *Twelfth Night: A Critical Reader*. Bloomsbury.

Flannery, T. (2005). *The Weather Makers: How Man is Changing the Climate and What it Means for Life on Earth*. Atlantic Monthly Press.

Frye, N. (1957). *Anatomy of Criticism: Four Essays*. Princeton University Press.

Gairola, R. K. (2016). *Homelandings: Postcolonial Diasporas and Transatlantic Belonging*. Rowman and Littlefield.

Gairola, R. K. (2019). Caliban's verses and curses: The dialect(ic)s of subordination in Shakespeare's *The Tempest* – from text to film. In R. Kumar and O. Prakash (Eds.), *Text, Context and Construction of Identity* (pp. 28–42). Cambridge Scholars Press.

Gallagher, S., & Lindgren, R. (2015). Enactive metaphors: Learning through full-body engagement. *Educational Psychology Review*, *27*(3), 391–404. https://doi.org/10.1007/s10648-015-9327-1

Gibbs Jr, R. W. (2019). Metaphor as dynamical–ecological performance. *Metaphor and Symbol*, *34*(1), 33–44. https://doi.org/10.1080/10926488.2019.1591713

Hamilton, J. M. (2017). *This Contentious Storm: An Ecocritical and Performance History of King Lear*. Bloomsbury.

Hansen, C. (2019). 'Tropic of Shakespeare: What studying *Macbeth* in Queensland could teach us about place and shipwrecks.' *The Conversation*, 11 October 2019.

Hansen, C. (2023a). Shakespeare, climate change and the blue humanities: Imagining an oceanic education. In P. Bickley & J. Stevens (Eds.), *Shakespeare and Education: Representations, Intersections and Appropriations*, pp. 190–199. Routledge.

Hansen, C. (2023b). *Shakespeare and Place-Based Learning*. Cambridge University Press.

Harari, Y. N. (2015). *Sapiens: A Brief History of Humankind*. Translated by John Purcell and Haim Watzman. Harper.

Hardy-Vallée, B., & Payette, N. (2008). Introduction. In B. Hardy-Vallée (Ed.), *Beyond the Brain: Embodied, Situated and Distributed Cognition* (pp. 1–12). Cambridge Scholars Press.

Khan, J. U. (2019). History (and/or historicity) of ecocriticism and ecocritical history: An introductory overview. *International Journal of Linguistics, Literature and Translation*, *2*(4), 89–102. https://www.al-kindipublisher.com/index.php/ijllt/article/view/1256

Lines, W. J. (1999). *Taming the Great South Land: A History of the Conquest of Nature in Australia*. University of Georgia Press.

Lovelock, J. (2006). *The Revenge of Gaia: Earth's Climate in Crisis and the Fate of Humanity*. New York: Basic Books.

Ma Rhea, Z. (2017). *Frontiers of Taste: Food Sovereignty, Sustainability, and Indigenous–Settler Relations in Australia*. Springer Nature.

Martin, R. (2015). *Shakespeare and Ecology*. Oxford University Press.
Mentz, S. (2009). *At the Bottom of Shakespeare's Ocean*. Bloomsbury.
Mentz, S. (2018). Green comedy: Shakespeare and ecology. In H. Hirschfeld (Ed.), *The Oxford Handbook of Shakespearean Comedy* (pp. 250–262). Oxford University Press.
Mentz, S. (2019). Shakespeare and the blue humanities. *Studies in English Literature, 1500–1900*, *59*(2), 383–392. https://doi.org/10.1353/sel.2019.0018
Minton, G. [Dir.] (2023). *Salt Waves Fresh*. Montana InSite Theatre and James Cook University.
Nichols, W. J. (2014). *Blue Mind*. Little, Brown and Company.
Nunn, T. [Dir.]. (1996). *Twelfth Night* [Film]. Fine Line Features.
Proudfoot, R., Thomson, A., Kastan, D. S., & Woudhuysen, H. R. (Eds.). (2013). *The Arden Shakespeare Complete Works*. Bloomsbury Arden Shakespeare.
Pyne, S. J. (2016). *Burning Bush: A Fire History of Australia*. University of Washington Press.
Smith, P. J. (2014) Goodbye wave: Trevor Nunn's *Twelfth Night* – A maritime valediction. *Shakespeare*, *14*(4), 390–398. https://doi.org/10.1080/17450918.2017.1381641
State Library of NSW (SLNSW) Paddle steamer 55 'Lancashire Lass' and barge with 1,158 bales of wool – Wilcannia, NSWc 1905'. Accessed 20 May 2023. https://www.sl.nsw.gov.au/collection-items/paddle-steamer-55-lancashire-lass-and-barge-1158-bales-wool-wilcannia-nsw
Thackway, R., Brown, V., March, D., & Harris, J. A. (2018). Land use planning as a collective learning spiral: The case of regenerative landscape policy and practice. In R. Thackway (Ed.), *Land Use in Australia: Past, Present and Future* (pp. 263–278). Australian National University.
Williams, K. [Dir.]. (2022). *The Tempest*. Sydney Theatre Company.
Zeunert, J. (2018). Challenges in agricultural sustainability and resilience: Towards regenerative practice. In J. Zeunert & T. Waterman (Eds.), *Routledge Handbook of Landscape and Food* (pp. 231–252). Routledge.
Zeunert, J. (2024). Bridging rural and urban disconnections: Spatial graphic explorations of Australia's livestock environments. In R. Pernice, B. Chen, & L. Cardellicchio (Eds.), *Urban Regeneration and Rural Revitalization in Australia and China*. Routledge.
Zeunert, J., & Freestone, R. (2022). From rural lands to agribusiness precincts: Agriculture in metropolitan Sydney 1948–2018. In C. Brisotto & F. Lemes de Oliveira (Eds.), *Re-imaging Resilient Food Environments – Perspectives from Planning History* (pp. 247–272). Springer.

INDEX

Aboriginal and Indigenous peoples 96, 117
Activism 7, 15, 95, 127, 135, 139–142, 190
Aesthetics 83, 94, 103, 112, 113, 124, 199, 203
Antarctic 36–38, 163, 165, 166, 167
Anthropocene 8, 93–95, 164
Anthropogenic 111–113, 118–120, 198, 199, 205, 208
Arts (contemporary) 14, 93–94, 105
Arts (discipline) 4, 6, 13, 14, 16, 53, 96, 112, 113, 163, 203
Australian blue humanities 1–5, 9, 13, 213, 219

Bleaching (coral) 112, 113, 203, 204
Blue Country 23–29
Blue humanities 4, 13, 42, 46, 52–54, 71, 79, 82, 85, 94, 95, 105, 111–118, 120, 121, 168, 212, 213
Blue Identity 13, 24, 29
Blue Shakespeare 212–217, 219

Capitalism 93–94, 118, 189
Climate change 3–7, 12, 16, 95, 96, 105, 111–113, 117–123, 128–135, 143, 189, 199, 201, 206, 217
Colonial (theory) 23–28, 50, 70, 79, 83, 96–99, 102–107, 151, 118, 190, 194
The Coral Battleground 117, 135, 137, 139, 141, 142
Coral Sea 198
Corals (animal) 63, 112, 119, 135, 198, 200–204

DeLoughrey, Elizabeth 102
Digital blue 124, 125, 127, 131
Digital solastalgia 123, 129–131

Ecological Imperialism 117, 118, 119
Ecological systems 70, 93, 112, 129, 130, 152, 157, 162, 164, 166, 203, 212, 213, 216, 219
Entanglements 5–8, 15, 16, 25, 46–54, 118, 150, 152, 159, 164, 172, 175, 179, 180, 182, 190, 194, 195, 199, 211
Epistemology 93, 101, 104, 179, 191

First Nations peoples 96–99

Governance 167–169
Great Barrier Reef 7–10, 52, 67, 111–117, 134, 172–175, 199, 203

Haraway, Donna 93
History (Environmental) 46, 50–52, 121
Humanities (Environmental) 1, 7–11, 23, 46, 58, 69, 93, 111, 123, 134, 149, 162, 172, 187, 198, 211–212

Ibis 149–159
Identity 98, 99, 128, 131, 188, 189, 190, 191, 193

Indigenous perspectives 10–12, 14, 29, 48, 50–54, 69–70, 84, 93–96, 104, 105, 113–120, 137, 187–191, 193, 199, 212, 217

Kinship 27, 59
Knowledge systems 23–25, 105, 112–113, 117, 120, 217

Literature 142–145
Littoral 27, 99

McCalman, Iain 50
Memory (the role of) 98, 188, 190–195
Mentz, Steve 4–9, 11–16, 52, 53, 60, 82, 211–214

Narrative (of) 51, 94, 118, 151
Ninney Rise 114–117

Ocean identity 128
Oceanic 25, 26, 28, 36, 40, 47, 53, 61, 63, 89, 94–99, 112
Oceanic regions 23, 36
Oceanic thinking (exhibition) 93–100, 103, 104
Oppermann, Serpil 3, 5, 6, 7, 8, 14, 15, 16, 61, 93, 118, 164, 168, 175

Polynesian 192, 212–214

Queensland 28, 46–52, 58, 84, 94–96, 99, 168, 172, 193–199, 205, 208

Rainforests 105, 149, 153, 159
Reef 64, 114, 123, 125, 136; *see also* Great Barrier Reef
Restoration (and adaptation) 119–120

Sanitary citizenship 71, 78, 79
Sea country 9–13, 135–138
Settler (colonialism) 27, 48, 49, 51, 52, 102, 104, 187, 191–194
Settler (concepts) 23, 70–71, 193
Shakespeare (blue) 212–218
Submarine world 85–91

Terraqueous 5–8, 36, 38, 41, 195
Tidalectic curating 94–96
Torres Strait Islander peoples 48, 69, 74, 104

Underwater/undersea 85–91, 100–103, 176–182

Villiers, Alan 13, 35, 36, 40

Water 24, 31, 46–49, 51–64, 69–78, 97, 104, 105, 118, 124, 125, 150–158, 164–166, 172–179, 192–199, 216–218
Water bodies 54, 63, 126
Water systems 111, 112
White settlers (Australia) 58, 71, 79, 117, 151
Wright, Judith 2, 114, 116

For Product Safety Concerns and Information please contact our EU
representative GPSR@taylorandfrancis.com
Taylor & Francis Verlag GmbH, Kaufingerstraße 24, 80331 München, Germany

www.ingramcontent.com/pod-product-compliance
Ingram Content Group UK Ltd.
Pitfield, Milton Keynes, MK11 3LW, UK
UKHW020009200926
475702UK00015B/2560

* 9 7 8 1 0 3 2 4 3 0 4 5 4 *